Introductory Chemistry

Introductory Chemistry

Edited by
Rylan Hayes

Larsen & Keller
www.larsen-keller.com

Introductory Chemistry
Edited by Rylan Hayes
ISBN: 978-1-63549-064-0 (Hardback)

⊟ Larsen & Keller

Published by Larsen and Keller Education,
5 Penn Plaza,
19th Floor,
New York, NY 10001, USA

Cataloging-in-Publication Data

Introductory chemistry / edited by Rylan Hayes.
        p. cm.
Includes bibliographical references and index.
ISBN 978-1-63549-064-0
    1. Chemistry. 2. Physical sciences. I. Hayes, Rylan.
QD31.3 .I58 2017
540--dc23

The publisher's policy is to use permanent paper from mills that operate a sustainable forestry policy. Furthermore, the publisher ensures that the text paper and cover boards used have met acceptable environmental accreditation standards.

Printed and bound in the United States of America.

For more information regarding Larsen and Keller Education and its products, please visit the publisher's website www.larsen-keller.com

Table of Contents

Preface

Chemistry is a vast field in the world of science. It is the study of the properties, structure, composition and changes of matter. It includes examining the properties and chemical bonds of atoms. It is concerned with the reactions which transform one element into another. This book aims to shed light on some of the unexplored aspects of chemistry. It is a valuable compilation of topics, ranging from the basic to the most complex theories and principles in this field. Different approaches, evaluations and methodologies and advanced studies on chemistry have been included in the text. As this field is emerging at a rapid pace, the contents of this book will help the readers understand the modern concepts and applications of the subject.

To facilitate a deeper understanding of the contents of this book a short introduction of every chapter is written below:

Chapter 1- Chemistry is the branch of science that is concerned with substances, and their properties and reactions. It has topics such as atoms, chemical compounds and the general properties of substances. This chapter on chemistry offers an insightful focus, keeping in mind the complex subject matter.

Chapter 2- The definition of chemistry has been constantly evolving and with the discovery of new elements and compounds, chemistry has become a dynamic field of study. This chapter is concerned with the history of chemistry and elucidates the evolution of chemistry.

Chapter 3- Matter and atoms are the building blocks of chemistry. The following chapter unfolds its vital aspects in a critical yet systematic manner. It strategically encompasses and incorporates the major components and key concepts of chemistry, providing a complete understanding.

Chapter 4- Chemistry is an interdisciplinary subject. This content will provide a glimpse of related fields of chemistry briefly. Biochemistry, physical chemistry and organic chemistry are some of the various branches of chemistry that are explained in this chapter. It provides a plethora of interdisciplinary topics for better comprehension of chemistry.

Chapter 5- Quantum chemistry is a branch of chemistry that deals with the application of quantum mechanics in experiments and physical models of chemical systems. Quantum chemistry is an emerging field of study and the following chapter will not only provide an overview, but it will also inquire deep into the diverse topics related to it.

Chapter 6- This chapter elucidates the crucial theories and principles on chemical compounds. A chemical compound is an entity consisting of two or more atoms; there are four types of compounds, depending on how the constituents are held together. The topics discussed are of great importance to broaden the existing knowledge on chemistry.

I would like to share the credit of this book with my editorial team who worked tirelessly on this book. I owe the completion of this book to the never-ending support of my family, who supported me throughout the project.

Editor

Introduction to Chemistry

Chemistry is the branch of science that is concerned with substances, and their properties and reactions. It has topics such as atoms, chemical compounds and the general properties of substances. This chapter on chemistry offers an insightful focus, keeping in mind the complex subject matter.

Solutions of substances in reagent bottles, including ammonium hydroxide and nitric acid, illuminated in different colors

Chemistry is sometimes called the central science because it bridges other natural sciences, including physics, geology and biology.

Scholars disagree about the etymology of the word *chemistry*. The history of chemistry can be traced to alchemy, which had been practiced for several millennia in various parts of the world.

Etymology

The word *chemistry* comes from the word *alchemy* which was an earlier set of practices that encompassed elements of chemistry, metallurgy, philosophy, astrology, astronomy, mysticism and medicine. It is often seen as linked to the quest to turn lead or another common starting material into gold. Alchemy, which was practiced around 330, is the study of the composition of waters, movement, growth, embodying, disembodying, drawing the spirits from bodies and bonding the spirits within bodies (Zosimos). An alchemist was called a 'chemist' in popular speech, and later the suffix "-ry" was added to this to describe the art of the chemist as "chemistry".

The word *alchemy* in turn is derived from the Arabic word *al-kīmīā*. In origin, the term is borrowed from the Greek χημία or χημεία. This may have Egyptian origins since *al-kīmīā* is derived from the Greek χημία, which is in turn derived from the word Chemi or Kimi, which is the ancient name of Egypt in Egyptian. Alternately, *al-kīmīā* may derive from χημεία, meaning "cast together."

Definition

In retrospect, the definition of chemistry has changed over time, as new discoveries and theories add to the functionality of the science. The term "chymistry", in the view of noted scientist Robert Boyle in 1661, meant the subject of the material principles of mixed bodies. In 1663 the chemist Christopher Glaser described "chymistry" as a scientific art, by which one learns to dissolve bodies, and draw from them the different substances on their composition, and how to unite them again, and exalt them to a higher perfection.

The 1730 definition of the word "chemistry", as used by Georg Ernst Stahl, meant the art of resolving mixed, compound, or aggregate bodies into their principles; and of composing such bodies from those principles. In 1837, Jean-Baptiste Dumas considered the word "chemistry" to refer to the science concerned with the laws and effects of molecular forces. This definition further evolved until, in 1947, it came to mean the science of substances: their structure, their properties, and the reactions that change them into other substances - a characterization accepted by Linus Pauling. More recently, in 1998, Professor Raymond Chang broadened the definition of "chemistry" to mean the study of matter and the changes it undergoes.

History

Democritus' atomist philosophy was later adopted by Epicurus (341–270 BCE).

Early civilizations, such as the Egyptians Babylonians, Indians amassed practical knowledge concerning the arts of metallurgy, pottery and dyes, but didn't develop a systematic theory.

A basic chemical hypothesis first emerged in Classical Greece with the theory of four elements as propounded definitively by Aristotle stating that that fire, air, earth and water were the fundamental elements from which everything is formed as a combination. Greek atomism dates back to 440 BC, arising in works by philosophers such as Democritus and Epicurus. In 50 BC, the Roman philosopher Lucretius expanded upon the theory in his book *De rerum natura* (On The Nature of Things). Unlike modern concepts of science, Greek atomism was purely philosophical in nature, with little concern for empirical observations and no concern for chemical experiments.

In the Hellenistic world the art of alchemy first proliferated, mingling magic and occultism into the study of natural substances with the ultimate goal of transmuting elements into gold and discovering the elixir of eternal life. Alchemy was discovered and practised widely throughout the Arab world after the Muslim conquests, and from there, diffused into medieval and Renaissance Europe through Latin translations.

Chemistry as Science

Under the influence of the new empirical methods propounded by Sir Francis Bacon and others, a group of chemists at Oxford, Robert Boyle, Robert Hooke and John Mayow began to reshape the old alchemical traditions into a scientific discipline. Boyle in particular is regarded as the founding father of chemistry due to his most important work, the classic chemistry text *The Sceptical Chymist* where the differentiation is made between the claims of alchemy and the empirical scientific discoveries of the new chemistry. He formulated Boyle's law, rejected the classical "four elements" and proposed a mechanistic alternative of atoms and chemical reactions that could be subject to rigorous experiment.

Antoine-Laurent de Lavoisier is considered the "Father of Modern Chemistry".

The theory of phlogiston (a substance at the root of all combustion) was propounded by the German Georg Ernst Stahl in the early 18th century and was only overturned by the end of the century by the French chemist Antoine Lavoisier, the chemical analogue of Newton in physics; who did more than any other to establish the new science on proper theoretical footing, by elucidating the principle of conservation of mass and developing a new system of chemical nomenclature used to this day.

Prior to his work, though, many important discoveries had been made, specifically relating to the nature of 'air' which was discovered to be composed of many different gases. The Scottish chemist Joseph Black (the first experimental chemist) and the Dutchman J. B. van Helmont discovered carbon dioxide, or what Black called 'fixed air' in 1754; Henry Cavendish discovered hydrogen and elucidated its properties and Joseph Priestley and, independently, Carl Wilhelm Scheele isolated pure oxygen.

In his periodic table, Dmitri Mendeleev predicted the existence of 7 new elements, and placed all 60 elements known at the time in their correct places.

English scientist John Dalton proposed the modern theory of atoms; that all substances are composed of indivisible 'atoms' of matter and that different atoms have varying atomic weights.

The development of the electrochemical theory of chemical combinations occurred in the early 19th century as the result of the work of two scientists in particular, J. J. Berzelius and Humphry Davy, made possible by the prior invention of the voltaic pile by Alessandro Volta. Davy discovered nine new elements including the alkali metals by extracting them from their oxides with electric current.

British William Prout first proposed ordering all the elements by their atomic weight as all atoms had a weight that was an exact multiple of the atomic weight of hydrogen. J. A. R. Newlands devised an early table of elements, which was then developed into the modern periodic table of elements in the 1860s by Dmitri Mendeleev and independently by several other scientists including Julius Lothar Meyer. The inert gases, later called the noble gases were discovered by William Ramsay in collaboration with Lord Rayleigh at the end of the century, thereby filling in the basic structure of the table.

Organic chemistry was developed by Justus von Liebig and others, following Friedrich Wöhler's synthesis of urea which proved that living organisms were, in theory, reducible to chemistry. Other crucial 19th century advances were; an understanding of valence bonding (Edward Frankland in 1852) and the application of thermodynamics to chemistry (J. W. Gibbs and Svante Arrhenius in the 1870s).

Chemical Structure

At the turn of the twentieth century the theoretical underpinnings of chemistry were finally understood due to a series of remarkable discoveries that succeeded in probing and discovering the very nature of the internal structure of atoms. In 1897, J. J. Thomson of Cambridge University discovered the electron and soon after the French scientist Becquerel as well as the couple Pierre and Marie Curie investigated the phenomenon of radioactivity. In a series of pioneering scattering experiments Ernest Rutherford at the University of Manchester discovered the internal structure of the atom and the existence of the proton, classified and explained the different types of radioactivity and successfully transmuted the first element by bombarding nitrogen with alpha particles.

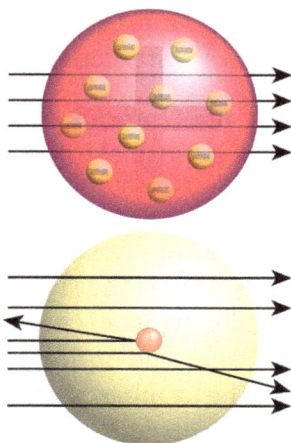

Top: Expected results: alpha particles passing through the plum pudding model of the atom undisturbed.
Bottom: Observed results: a small portion of the particles were deflected, indicating a small, concentrated charge.

His work on atomic structure was improved on by his students, the Danish physicist Niels Bohr and Henry Moseley. The electronic theory of chemical bonds and molecular orbitals was developed by the American scientists Linus Pauling and Gilbert N. Lewis.

The year 2011 was declared by the United Nations as the International Year of Chemistry. It was an initiative of the International Union of Pure and Applied Chemistry, and of the United Nations Educational, Scientific, and Cultural Organization and involves chemical societies, academics, and institutions worldwide and relied on individual initiatives to organize local and regional activities.

Principles of Modern Chemistry

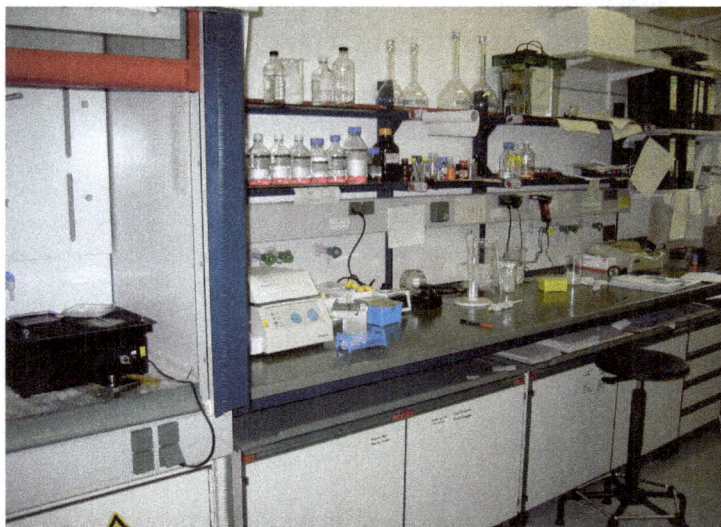

Laboratory, Institute of Biochemistry, University of Cologne.

The current model of atomic structure is the quantum mechanical model. Traditional chemistry starts with the study of elementary particles, atoms, molecules, substances, metals, crystals and other aggregates of matter. This matter can be studied in solid, liquid, or gas states, in isolation or in combination. The interactions, reactions and transformations that are studied in chemistry are usually the result of interactions between atoms, leading to rearrangements of the chemical bonds which hold atoms together. Such behaviors are studied in a chemistry laboratory.

The chemistry laboratory stereotypically uses various forms of laboratory glassware. However glassware is not central to chemistry, and a great deal of experimental (as well as applied/industrial) chemistry is done without it.

A chemical reaction is a transformation of some substances into one or more different substances. The basis of such a chemical transformation is the rearrangement of electrons in the chemical bonds between atoms. It can be symbolically depicted through a chemical equation, which usually involves atoms as subjects. The number of atoms on the left and the right in the equation for a chemical transformation is equal. (When the number of atoms on either side is unequal, the transformation is referred to as a nuclear reaction or radioactive decay.) The type of chemical reactions a substance may undergo and the energy changes that may accompany it are constrained by certain basic rules, known as chemical laws.

Energy and entropy considerations are invariably important in almost all chemical studies. Chemical substances are classified in terms of their structure, phase, as well as their chemical compositions. They can be analyzed using the tools of chemical analysis, e.g. spectroscopy and chromatography. Scientists engaged in chemical research are known as chemists. Most chemists specialize in one or more sub-disciplines. Several concepts are essential for the study of chemistry; some of them are:

Matter

In chemistry, matter is defined as anything that has rest mass and volume (it takes up space) and is made up of particles. The particles that make up matter have rest mass as well - not all particles have rest mass, such as the photon. Matter can be a pure chemical substance or a mixture of substances.

Atom

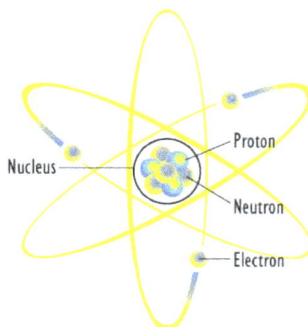

A diagram of an atom based on the Rutherford model

The atom is the basic unit of chemistry. It consists of a dense core called the atomic nucleus surrounded by a space called the electron cloud. The nucleus is made up of positively charged protons and uncharged neutrons (together called nucleons), while the electron cloud consists of negatively charged electrons which orbit the nucleus. In a neutral atom, the negatively charged electrons balance out the positive charge of the protons. The nucleus is dense; the mass of a nucleon is 1,836 times that of an electron, yet the radius of an atom is about 10,000 times that of its nucleus.

The atom is also the smallest entity that can be envisaged to retain the chemical properties of the element, such as electronegativity, ionization potential, preferred oxidation state(s), coordination number, and preferred types of bonds to form (e.g., metallic, ionic, covalent).

Element

Standard form of the periodic table of chemical elements. The colors represent different categories of elements

A chemical element is a pure substance which is composed of a single type of atom, characterized by its particular number of protons in the nuclei of its atoms, known as the atomic number and represented by the symbol Z. The mass number is the sum of the number of protons and neutrons in a nucleus. Although all the nuclei of all atoms belonging to one element will have the same atomic number, they may not necessarily have the same mass number; atoms of an element which have different mass numbers are known as isotopes. For example, all atoms with 6 protons in their nuclei are atoms of the chemical element carbon, but atoms of carbon may have mass numbers of 12 or 13.

The standard presentation of the chemical elements is in the periodic table, which orders elements by atomic number. The periodic table is arranged in groups, or columns, and periods, or rows. The periodic table is useful in identifying periodic trends.

Compound

Carbon dioxide (CO_2), an example of a chemical compound

A *compound* is a pure chemical substance composed of more than one element. The properties of a compound bear little similarity to those of its elements. The standard nomenclature of compounds is set by the International Union of Pure and Applied Chemistry (IUPAC). Organic compounds are named according to the organic nomenclature system. Inorganic compounds are named according to the inorganic nomenclature system. In addition the Chemical Abstracts Service has devised a method to index chemical substances. In this scheme each chemical substance is identifiable by a number known as its CAS registry number.

Molecule

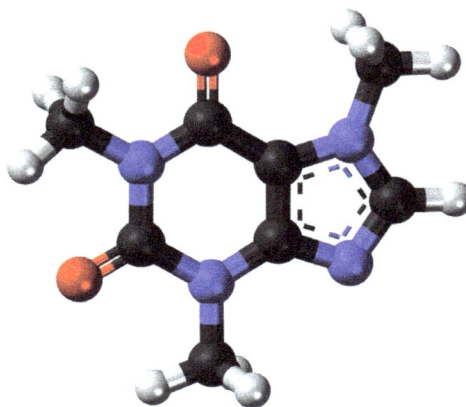

A ball-and-stick representation of the caffeine molecule ($C_8H_{10}N_4O_2$).

A *molecule* is the smallest indivisible portion of a pure chemical substance that has its unique set of chemical properties, that is, its potential to undergo a certain set of chemical reactions with other substances. However, this definition only works well for substances that are composed of molecules, which is not true of many substances. Molecules are typically a set of atoms bound together by covalent bonds, such that the structure is electrically neutral and all valence electrons are paired with other electrons either in bonds or in lone pairs.

Thus, molecules exist as electrically neutral units, unlike ions. When this rule is broken, giving the "molecule" a charge, the result is sometimes named a molecular ion or a polyatomic ion. However, the discrete and separate nature of the molecular concept usually requires that molecular ions be present only in well-separated form, such as a directed beam in a vacuum in a mass spectrometer. Charged polyatomic collections residing in solids (for example, common sulfate or nitrate ions) are generally not considered "molecules" in chemistry.

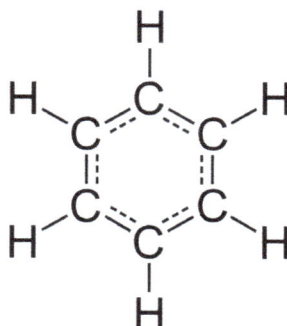

A 2-D skeletal model of a benzene molecule (C_6H_6)

The "inert" or noble gas elements (helium, neon, argon, krypton, xenon and radon) are composed of lone atoms as their smallest discrete unit, but the other isolated chemical elements consist of either molecules or networks of atoms bonded to each other in some way. Identifiable molecules compose familiar substances such as water, air, and many organic compounds like alcohol, sugar, gasoline, and the various pharmaceuticals.

However, not all substances or chemical compounds consist of discrete molecules, and indeed most of the solid substances that make up the solid crust, mantle, and core of the Earth are chemical compounds without molecules. These other types of substances, such as ionic compounds and network solids, are organized in such a way as to lack the existence of identifiable molecules *per se*. Instead, these substances are discussed in terms of formula units or unit cells as the smallest repeating structure within the substance. Examples of such substances are mineral salts (such as table salt), solids like carbon and diamond, metals, and familiar silica and silicate minerals such as quartz and granite.

One of the main characteristics of a molecule is its geometry often called its structure. While the structure of diatomic, triatomic or tetra atomic molecules may be trivial, (linear, angular pyramidal etc.) the structure of polyatomic molecules, that are constituted of more than six atoms (of several elements) can be crucial for its chemical nature.

Substance and Mixture

A chemical substance is a kind of matter with a definite composition and set of properties. A collection of substances is called a mixture. Examples of mixtures are air and alloys.

Mole and Amount of Substance

The mole is a unit of measurement that denotes an amount of substance (also called chemical amount). The mole is defined as the number of atoms found in exactly 0.012 kilogram (or 12 grams) of carbon-12, where the carbon-12 atoms are unbound, at rest and in their ground state. The number of entities per mole is known as the Avogadro constant, and is determined empirically to be approximately 6.022×10^{23} mol^{-1}. Molar concentration is the amount of a particular substance per volume of solution, and is commonly reported in moldm^{-3}.

Phase

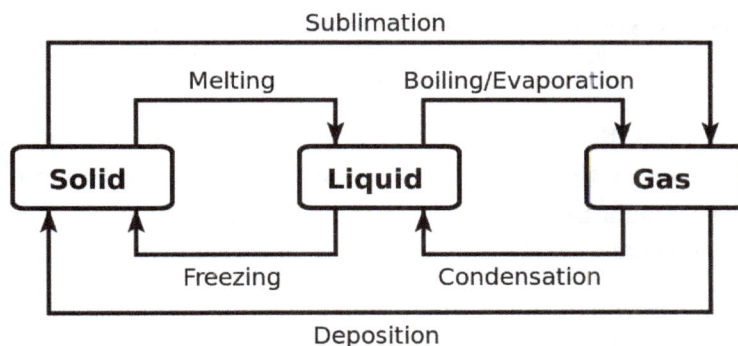

Example of phase changes

In addition to the specific chemical properties that distinguish different chemical classifications, chemicals can exist in several phases. For the most part, the chemical classifications are independent of these bulk phase classifications; however, some more exotic phases are incompatible with certain chemical properties. A *phase* is a set of states of a chemical system that have similar bulk structural properties, over a range of conditions, such as pressure or temperature.

Physical properties, such as density and refractive index tend to fall within values characteristic of the phase. The phase of matter is defined by the *phase transition*, which is when energy put into or taken out of the system goes into rearranging the structure of the system, instead of changing the bulk conditions.

Sometimes the distinction between phases can be continuous instead of having a discrete boundary, in this case the matter is considered to be in a supercritical state. When three states meet based on the conditions, it is known as a triple point and since this is invariant, it is a convenient way to define a set of conditions.

The most familiar examples of phases are solids, liquids, and gases. Many substances exhibit multiple solid phases. For example, there are three phases of solid iron (alpha, gamma, and delta) that vary based on temperature and pressure. A principal difference between solid phases is the crystal structure, or arrangement, of the atoms. Another phase commonly encountered in the study of chemistry is the *aqueous* phase, which is the state of substances dissolved in aqueous solution (that is, in water).

Less familiar phases include plasmas, Bose–Einstein condensates and fermionic condensates and the paramagnetic and ferromagnetic phases of magnetic materials. While most familiar phases deal with three-dimensional systems, it is also possible to define analogs in two-dimensional systems, which has received attention for its relevance to systems in biology.

Bonding

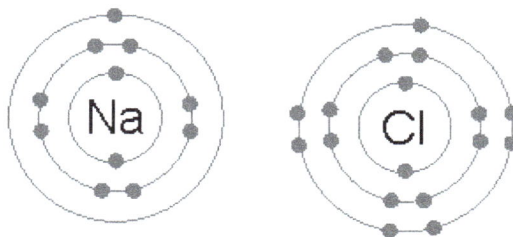

Process of ionic bonding between sodium (Na) and chlorine (Cl) to form sodium chloride, or common table salt. Ionic bonding involves one atom taking valence electrons from another (as opposed to sharing, which occurs in covalent bonding)

Atoms sticking together in molecules or crystals are said to be bonded with one another. A chemical bond may be visualized as the multipole balance between the positive charges in the nuclei and the negative charges oscillating about them. More than simple attraction and repulsion, the energies and distributions characterize the availability of an electron to bond to another atom.

A chemical bond can be a covalent bond, an ionic bond, a hydrogen bond or just because of Van der Waals force. Each of these kinds of bonds is ascribed to some potential. These potentials create the interactions which hold atoms together in molecules or crystals. In many simple compounds, valence bond theory, the Valence Shell Electron Pair Repulsion model (VSEPR), and the concept of oxidation number can be used to explain molecular structure and composition.

An ionic bond is formed when a metal loses one or more of its electrons, becoming a positively charged cation, and the electrons are then gained by the non-metal atom, becoming a negatively charged anion. The two oppositely charged ions attract one another, and the ionic bond is the electrostatic force of attraction between them. For example, sodium (Na), a metal, loses one electron to become an Na^+ cation while chlorine (Cl), a non-metal, gains this electron to become Cl^-. The ions are held together due to electrostatic attraction, and that compound sodium chloride (NaCl), or common table salt, is formed.

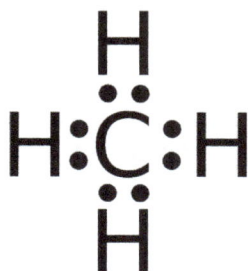

$$H:\overset{\displaystyle H}{\underset{\displaystyle H}{C}}:H$$

In the methane molecule (CH_4), the carbon atom shares a pair of valence electrons with each of the four hydrogen atoms. Thus, the octet rule is satisfied for C-atom (it has eight electrons in its valence shell) and the duet rule is satisfied for the H-atoms (they have two electrons in their valence shells).

In a covalent bond, one or more pairs of valence electrons are shared by two atoms: the resulting electrically neutral group of bonded atoms is termed a molecule. Atoms will share valence electrons in such a way as to create a noble gas electron configuration (eight electrons in their outermost shell) for each atom. Atoms that tend to combine in such a way that they each have eight electrons in their valence shell are said to follow the octet rule. However, some elements like hydrogen and lithium need only two electrons in their outermost shell to attain this stable configuration; these atoms are said to follow the *duet rule*, and in this way they are reaching the electron configuration of the noble gas helium, which has two electrons in its outer shell.

Similarly, theories from classical physics can be used to predict many ionic structures. With more complicated compounds, such as metal complexes, valence bond theory is less applicable and alternative approaches, such as the molecular orbital theory, are generally used. See diagram on electronic orbitals.

Energy

In the context of chemistry, energy is an attribute of a substance as a consequence of its atomic, molecular or aggregate structure. Since a chemical transformation is accompanied by a change in one or more of these kinds of structures, it is invariably accompanied by an increase or decrease of energy of the substances involved. Some energy is transferred between the surroundings and the reactants of the reaction in the form of heat or light; thus the products of a reaction may have more or less energy than the reactants.

A reaction is said to be exergonic if the final state is lower on the energy scale than the initial state; in the case of endergonic reactions the situation is the reverse. A reaction is said to be exothermic if the reaction releases heat to the surroundings; in the case of endothermic reactions, the reaction absorbs heat from the surroundings.

Chemical reactions are invariably not possible unless the reactants surmount an energy barrier known as the activation energy. The *speed* of a chemical reaction (at given temperature T) is related to the activation energy E, by the Boltzmann's population factor $e^{-E/kT}$ - that is the probability of a molecule to have energy greater than or equal to E at the given temperature T. This exponential dependence of a reaction rate on temperature is known as the Arrhenius equation. The activation energy necessary for a chemical reaction to occur can be in the form of heat, light, electricity or mechanical force in the form of ultrasound.

A related concept free energy, which also incorporates entropy considerations, is a very useful means for predicting the feasibility of a reaction and determining the state of equilibrium of a chemical reaction, in chemical thermodynamics. A reaction is feasible only if the total change in the Gibbs free energy is negative, $\Delta G \leq 0$; if it is equal to zero the chemical reaction is said to be at equilibrium.

There exist only limited possible states of energy for electrons, atoms and molecules. These are determined by the rules of quantum mechanics, which require quantization of energy of a bound system. The atoms/molecules in a higher energy state are said to be excited. The molecules/atoms of substance in an excited energy state are often much more reactive; that is, more amenable to chemical reactions.

The phase of a substance is invariably determined by its energy and the energy of its surroundings. When the intermolecular forces of a substance are such that the energy of the surroundings is not sufficient to overcome them, it occurs in a more ordered phase like liquid or solid as is the case with water (H_2O); a liquid at room temperature because its molecules are bound by hydrogen bonds. Whereas hydrogen sulfide (H_2S) is a gas at room temperature and standard pressure, as its molecules are bound by weaker dipole-dipole interactions.

The transfer of energy from one chemical substance to another depends on the *size* of energy quanta emitted from one substance. However, heat energy is often transferred more easily from almost any substance to another because the phonons responsible for vibrational and rotational energy levels in a substance have much less energy than photons invoked for the electronic energy transfer. Thus, because vibrational and rotational energy levels are more closely spaced than electronic energy levels, heat is more easily transferred between substances relative to light or other forms of electronic energy. For example, ultraviolet electromagnetic radiation is not transferred with as much efficacy from one substance to another as thermal or electrical energy.

The existence of characteristic energy levels for different chemical substances is useful for their identification by the analysis of spectral lines. Different kinds of spectra are often used in chemical spectroscopy, e.g. IR, microwave, NMR, ESR, etc. Spectroscopy is also used to identify the composition of remote objects - like stars and distant galaxies - by analyzing their radiation spectra.

Emission spectrum of iron

The term chemical energy is often used to indicate the potential of a chemical substance to undergo a transformation through a chemical reaction or to transform other chemical substances.

Reaction

During chemical reactions, bonds between atoms break and form, resulting in different substances with different properties. In a blast furnace, iron oxide, a compound, reacts with carbon monoxide to form iron, one of the chemical elements, and carbon dioxide.

When a chemical substance is transformed as a result of its interaction with another substance or with energy, a chemical reaction is said to have occurred. A *chemical reaction* is therefore a concept related to the "reaction" of a substance when it comes in close contact with another, whether as a mixture or a solution; exposure to some form of energy, or both. It results in some energy exchange between the constituents of the reaction as well as with the system environment, which may be designed vessels—often laboratory glassware.

Chemical reactions can result in the formation or dissociation of molecules, that is, molecules breaking apart to form two or more smaller molecules, or rearrangement of atoms within or across molecules. Chemical reactions usually involve the making or breaking of chemical bonds. Oxidation, reduction, dissociation, acid-base neutralization and molecular rearrangement are some of the commonly used kinds of chemical reactions.

A chemical reaction can be symbolically depicted through a chemical equation. While in a non-nuclear chemical reaction the number and kind of atoms on both sides of the equation are equal, for a nuclear reaction this holds true only for the nuclear particles viz. protons and neutrons.

The sequence of steps in which the reorganization of chemical bonds may be taking place in the course of a chemical reaction is called its mechanism. A chemical reaction can be envisioned to take place in a number of steps, each of which may have a different speed. Many reaction intermediates with variable stability can thus be envisaged during the course of a reaction. Reaction mechanisms are proposed to explain the kinetics and the relative product mix of a reaction. Many physical chemists specialize in exploring and proposing the mechanisms of various chemical reactions. Several empirical rules, like the Woodward–Hoffmann rules often come in handy while proposing a mechanism for a chemical reaction.

According to the IUPAC gold book, a chemical reaction is "a process that results in the interconversion of chemical species." Accordingly, a chemical reaction may be an elementary reaction or a stepwise reaction. An additional caveat is made, in that this definition includes cases where the interconversion of conformers is experimentally observable. Such detectable chemical reactions normally involve sets of molecular entities as indicated by this definition, but it is often conceptually convenient to use the term also for changes involving single molecular entities (i.e. 'microscopic chemical events').

Ions and Salts

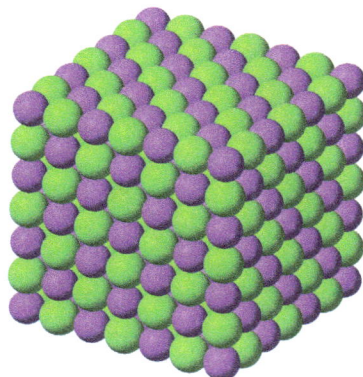

The crystal lattice structure of potassium chloride (KCl), a salt which is formed due to the attraction of K^+ cations and Cl^- anions. Note how the overall charge of the ionic compound is zero.

An *ion* is a charged species, an atom or a molecule, that has lost or gained one or more electrons. When an atom loses an electron and thus has more protons than electrons, the atom is a positively charged ion or cation. When an atom gains an electron and thus has more electrons than protons, the atom is a negatively charged ion or anion. Cations and anions can form a crystalline lattice of neutral salts, such as the Na^+ and Cl^- ions forming sodium chloride, or NaCl. Examples of polyatomic ions that do not split up during acid-base reactions are hydroxide (OH^-) and phosphate (PO_4^{3-}).

Plasma is composed of gaseous matter that has been completely ionized, usually through high temperature.

Acidity and Basicity

A substance can often be classified as an acid or a base. There are several different theories which explain acid-base behavior. The simplest is Arrhenius theory, which states than an acid is a substance that produces hydronium ions when it is dissolved in water, and a base is one that produces hydroxide ions when dissolved in water. According to Brønsted–Lowry acid–base theory, acids are substances that donate a positive hydrogen ion to another substance in a chemical reaction; by extension, a base is the substance which receives that hydrogen ion.

When hydrogen bromide (HBr), pictured, is dissolved in water, it forms the strong acid hydrobromic acid

A third common theory is Lewis acid-base theory, which is based on the formation of new chemical bonds. Lewis theory explains that an acid is a substance which is capable of accepting a pair of electrons from another substance during the process of bond formation, while a base is a substance which can provide a pair of electrons to form a new bond. According to this theory, the crucial things being exchanged are charges. There are several other ways in which a substance may be classified as an acid or a base, as is evident in the history of this concept.

Acid strength is commonly measured by two methods. One measurement, based on the Arrhenius definition of acidity, is pH, which is a measurement of the hydronium ion concentration in a solution, as expressed on a negative logarithmic scale. Thus, solutions that have a low pH have a high hydronium ion concentration, and can be said to be more acidic. The other measurement, based on the Brønsted–Lowry definition, is the acid dissociation constant (K_a), which measures the relative ability of a substance to act as an acid under the Brønsted–Lowry definition of an acid. That is, substances with a higher K_a are more likely to donate hydrogen ions in chemical reactions than those with lower K_a values.

Redox

Redox (*reduction-oxidation*) reactions include all chemical reactions in which atoms have their oxidation state changed by either gaining electrons (reduction) or losing electrons (oxidation). Substances that have the ability to oxidize other substances are said to be oxidative and are known as oxidizing agents, oxidants or oxidizers. An oxidant removes electrons from another substance. Similarly, substances that have the ability to reduce other substances are said to be reductive and are known as reducing agents, reductants, or reducers.

A reductant transfers electrons to another substance, and is thus oxidized itself. And because it "donates" electrons it is also called an electron donor. Oxidation and reduction properly refer to a change in oxidation number—the actual transfer of electrons may never occur. Thus, oxidation is better defined as an increase in oxidation number, and reduction as a decrease in oxidation number.

Equilibrium

Although the concept of equilibrium is widely used across sciences, in the context of chemistry, it arises whenever a number of different states of the chemical composition are possible, as for example, in a mixture of several chemical compounds that can react with one another, or when a substance can be present in more than one kind of phase.

A system of chemical substances at equilibrium, even though having an unchanging composition, is most often not static; molecules of the substances continue to react with one another thus giving rise to a dynamic equilibrium. Thus the concept describes the state in which the parameters such as chemical composition remain unchanged over time.

Chemical Laws

Chemical reactions are governed by certain laws, which have become fundamental concepts in chemistry. Some of them are:

- Avogadro's law

- Beer–Lambert law

- Boyle's law (1662, relating pressure and volume)

- Charles's law (1787, relating volume and temperature)

- Fick's laws of diffusion

- Gay-Lussac's law (1809, relating pressure and temperature)

- Le Chatelier's principle

- Henry's law

- Hess's law

- Law of conservation of energy leads to the important concepts of equilibrium, thermodynamics, and kinetics.

- Law of conservation of mass continues to be conserved in isolated systems, even in modern physics. However, special relativity shows that due to mass–energy equivalence, whenever non-material "energy" (heat, light, kinetic energy) is removed from a non-isolated system, some mass will be lost with it. High energy losses result in loss of weighable amounts of mass, an important topic in nuclear chemistry.

- Law of definite composition, although in many systems (notably biomacromolecules and minerals) the ratios tend to require large numbers, and are frequently represented as a fraction.

- Law of multiple proportions

- Raoult's law

Practice

Subdisciplines

Chemistry is typically divided into several major sub-disciplines. There are also several main cross-disciplinary and more specialized fields of chemistry.

- Analytical chemistry is the analysis of material samples to gain an understanding of their chemical composition and structure. Analytical chemistry incorporates standardized experimental methods in chemistry. These methods may be used in all subdisciplines of chemistry, excluding purely theoretical chemistry.

- Biochemistry is the study of the chemicals, chemical reactions and chemical interactions that take place in living organisms. Biochemistry and organic chemistry are closely related, as in medicinal chemistry or neurochemistry. Biochemistry is also associated with molecular biology and genetics.

- Inorganic chemistry is the study of the properties and reactions of inorganic compounds. The distinction between organic and inorganic disciplines is not absolute and there is much overlap, most importantly in the sub-discipline of organometallic chemistry.

- Materials chemistry is the preparation, characterization, and understanding of substances with a useful function. The field is a new breadth of study in graduate programs, and it integrates elements from all classical areas of chemistry with a focus on fundamental issues that are unique to materials. Primary systems of study include the chemistry of condensed phases (solids, liquids, polymers) and interfaces between different phases.

- Neurochemistry is the study of neurochemicals; including transmitters, peptides, proteins, lipids, sugars, and nucleic acids; their interactions, and the roles they play in forming, maintaining, and modifying the nervous system.

- Nuclear chemistry is the study of how subatomic particles come together and make nuclei. Modern Transmutation is a large component of nuclear chemistry, and the table of nuclides is an important result and tool for this field.

- Organic chemistry is the study of the structure, properties, composition, mechanisms, and reactions of organic compounds. An organic compound is defined as any compound based on a carbon skeleton.

- Physical chemistry is the study of the physical and fundamental basis of chemical systems and processes. In particular, the energetics and dynamics of such systems and processes are of interest to physical chemists. Important areas of study include chemical thermodynamics, chemical kinetics, electrochemistry, statistical mechanics, spectroscopy, and more recently, astrochemistry. Physical chemistry has large overlap with molecular physics. Physical chemistry involves the use of infinitesimal calculus in deriving equations. It is usually associated with quantum chemistry and theoretical chemistry. Physical chemistry is a distinct discipline from chemical physics, but again, there is very strong overlap.

- Theoretical chemistry is the study of chemistry via fundamental theoretical reasoning (usually within mathematics or physics). In particular the application of quantum mechanics to chemistry is called quantum chemistry. Since the end of the Second World War, the development of computers has allowed a systematic development of computational chemistry, which is the art of developing and applying computer programs for solving chemical problems. Theoretical chemistry has large overlap with (theoretical and experimental) condensed matter physics and molecular physics.

Other disciplines within chemistry are traditionally grouped by the type of matter being studied or the kind of study. These include inorganic chemistry, the study of inorganic matter; organic chemistry, the study of organic (carbon-based) matter; biochemistry, the study of substances found in biological organisms; physical chemistry, the study of chemical processes using physical concepts such as thermodynamics and quantum mechanics; and analytical chemistry, the analysis of material samples to gain an understanding of their chemical composition and structure. Many more specialized disciplines have emerged in recent years, e.g. neurochemistry the chemical study of the nervous system.

Other fields include agrochemistry, astrochemistry (and cosmochemistry), atmospheric chemistry, chemical engineering, chemical biology, chemo-informatics, electrochemistry, environmental chemistry, femtochemistry, flavor chemistry, flow chemistry, geochemistry, green chemistry, histochemistry, history of chemistry, hydrogenation chemistry, immunochemistry, marine chemistry, materials science, mathematical chemistry, mechanochemistry, medicinal chemistry, molecular biology, molecular mechanics, nanotechnology, natural product chemistry, oenology, organometallic chemistry, petrochemistry, pharmacology, photochemistry, physical organic chemistry, phytochemistry, polymer chemistry, radiochemistry, solid-state chemistry, sonochemistry, supramolecular chemistry, surface chemistry, synthetic chemistry, thermochemistry, and many others.

Chemical Industry

The chemical industry represents an important economic activity worldwide. The global top 50 chemical producers in 2013 had sales of US$980.5 billion with a profit margin of 10.3%.

Professional Societies

- American Chemical Society

- American Society for Neurochemistry

- Chemical Institute of Canada

- Chemical Society of Peru

- International Union of Pure and Applied Chemistry

- Royal Australian Chemical Institute

- Royal Netherlands Chemical Society

- Royal Society of Chemistry

- Society of Chemical Industry

- World Association of Theoretical and Computational Chemists
- List of chemistry societies

References

- Theodore L. Brown, H. Eugene Lemay, Bruce Edward Bursten, H. Lemay. Chemistry: The Central Science. Prentice Hall; 8 edition (1999). ISBN 0-13-010310-1.

- Chemistry occupies an intermediate position in a hierarchy of the sciences by reductive level between physics and biology. Carsten Reinhardt. Chemical Sciences in the 20th Century: Bridging Boundaries. Wiley-VCH, 2001. ISBN 3-527-30271-9.

- "Arabic alchemy", Georges C. Anawati, pp. 853–885 in Encyclopedia of the history of Arabic science, eds. Roshdi Rashed and Régis Morelon, London: Routledge, 1996, vol. 3, ISBN 0-415-12412-3.

- Glaser, Christopher (1663). Traite de la chymie. Paris. as found in: Kim, Mi Gyung (2003). Affinity, That Elusive Dream - A Genealogy of the Chemical Revolution. The MIT Press. ISBN 0-262-11273-6.

- Mi Gyung Kim (2003). Affinity, that Elusive Dream: A Genealogy of the Chemical Revolution. MIT Press. p. 440. ISBN 0-262-11273-6.

- Ihde, Aaron John (1984). The Development of Modern Chemistry. Courier Dover Publications. p. 164. ISBN 0-486-64235-6.

- Armstrong, James (2012). General, Organic, and Biochemistry: An Applied Approach. Brooks/Cole. p. 48. ISBN 978-0-534-49349-3.

- Chemistry 412 course notes. "A Brief History of the Development of Periodic Table". Western Oregon University. Retrieved July 20, 2015.

- Tullo, Alexander H. (28 July 2014). "C&EN's Global Top 50 Chemical Firms For 2014". Chemical & Engineering News (American Chemical Society). Retrieved 22 August 2014.

- "Ancients & Alchemists - Time line of achievement". Chemical Heritage Society. Archived from the original on 20 June 2010. Retrieved 23 March 2014.

- Winter, Mark. "WebElements: the periodic table on the web". The University of Sheffield. Archived from the original on January 4, 2014. Retrieved January 27, 2014.

- "International Year of Chemistry - The History of Chemistry". G.I.T. Laboratory Journal Europe. Feb 25, 2011. Retrieved March 12, 2013.

Evolution of Chemistry

The definition of chemistry has been constantly evolving and with the discovery of new elements and compounds, chemistry has become a dynamic field of study. This chapter is concerned with the history of chemistry and elucidates the evolution of chemistry.

History of Chemistry

The history of chemistry represents a time span from ancient history to the present. By 1000 BC, civilizations used technologies that would eventually form the basis of the various branches of chemistry. Examples include extracting metals from ores, making pottery and glazes, fermenting beer and wine, extracting chemicals from plants for medicine and perfume, rendering fat into soap, making glass, and making alloys like bronze.

Reihen	Gruppo I. — R'O	Gruppo II. — RO	Gruppo III. — R²O³	Gruppo IV. RH⁴ RO²	Gruppo V. RH³ R²O⁵	Gruppo VI. RH² RO³	Gruppo VII. RH R²O⁷	Gruppo VIII. — RO⁴
1	H=1							
2	Li=7	Be=9,4	B=11	C=12	N=14	O=16	F=19	
3	Na=23	Mg=24	Al=27,8	Si=28	P=31	S=32	Cl=35,5	
4	K=39	Ca=40	—=44	Ti=48	V=51	Cr=52	Mn=55	Fe=56, Co=59, Ni=59, Cu=63.
5	(Cu=63)	Zn=65	—=68	—=72	As=75	Se=78	Br=80	
6	Rb=85	Sr=87	?Yt=88	Zr=90	Nb=94	Mo=96	—=100	Ru=104, Rh=104, Pd=106, Ag=108
7	(Ag=108)	Cd=112	In=113	Sn=118	Sb=122	Te=125	J=127	
8	Cs=133	Ba=137	?Di=138	?Ce=140	—	—	—	— — —
9	(—)	—	—	—	—	—	—	
10	—	—	?Er=178	?La=180	Ta=182	W=184	—	Os=195, Ir=197, Pt=198, Au=199.
11	(Au=199)	Hg=200	Tl=204	Pb=207	Bi=208	—	—	
12	—	—	—	Th=231	—	U=240	—	— — —

The 1871 periodic table constructed by Dmitri Mendeleev. The periodic table is one of the most potent icons in science, lying at the core of chemistry and embodying the most fundamental principles of the field.

The protoscience of chemistry, alchemy, was unsuccessful in explaining the nature of matter and its transformations. However, by performing experiments and recording the results, alchemists set the stage for modern chemistry. The distinction began to emerge when a clear differentiation was made between chemistry and alchemy by Robert Boyle in his work *The Sceptical Chymist* (1661). While both alchemy and chemistry are concerned with matter and its transformations, chemists are seen as applying scientific method to their work.

Chemistry is considered to have become an established science with the work of Antoine Lavoisier, who developed a law of conservation of mass that demanded careful measurement and quantitative observations of chemical phenomena. The history of chemistry is intertwined with the history of thermodynamics, especially through the work of Willard Gibbs.

Ancient History

Early Metallurgy

The earliest recorded metal employed by humans seems to be gold which can be found free or "native". Small amounts of natural gold have been found in Spanish caves used during the late Paleolithic period, *c.* 40,000 BC.

Silver, copper, tin and meteoric iron can also be found native, allowing a limited amount of metal-working in ancient cultures. Egyptian weapons made from meteoric iron in about 3000 BC were highly prized as "Daggers from Heaven".

Arguably the first chemical reaction used in a controlled manner was fire. However, for millennia fire was seen simply as a mystical force that could transform one substance into another (burning wood, or boiling water) while producing heat and light. Fire affected many aspects of early societies. These ranged from the simplest facets of everyday life, such as cooking and habitat lighting, to more advanced technologies, such as pottery, bricks, and melting of metals to make tools.

It was fire that led to the discovery of glass and the purification of metals which in turn gave way to the rise of metallurgy. During the early stages of metallurgy, methods of purification of metals were sought, and gold, known in ancient Egypt as early as 2900 BC, became a precious metal.

Bronze Age

Certain metals can be recovered from their ores by simply heating the rocks in a fire: notably tin, lead and (at a higher temperature) copper, a process known as smelting. The first evidence of this extractive metallurgy dates from the 5th and 6th millennium BC, and was found in the archaeological sites of Majdanpek, Yarmovac and Plocnik, all three in Serbia. To date, the earliest copper smelting is found at the Belovode site, these examples include a copper axe from 5500 BC belonging to the Vinča culture. Other signs of early metals are found from the third millennium BC in places like Palmela (Portugal), Los Millares (Spain), and Stonehenge (United Kingdom). However, as often happens with the study of prehistoric times, the ultimate beginnings cannot be clearly defined and new discoveries are continuous and ongoing.

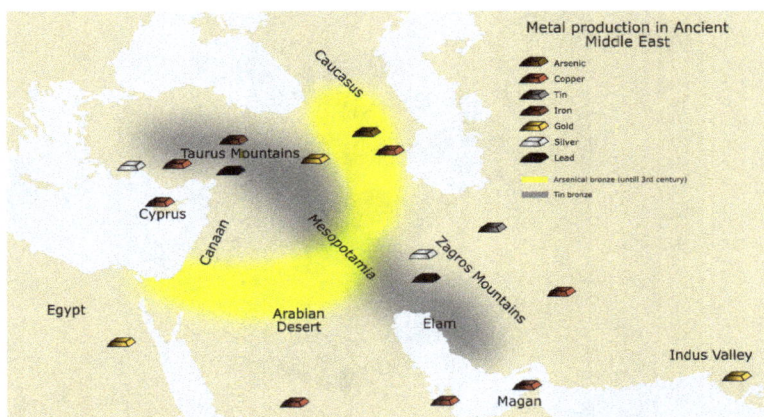

Mining areas of the ancient Middle East. Boxes colors: arsenic is in brown, copper in red, tin in grey, iron in reddish brown, gold in yellow, silver in white and lead in black. Yellow area stands for arsenic bronze, while grey area stands for tin bronze.

These first metals were single ones or as found. By combining copper and tin, a superior metal could be made, an alloy called bronze, a major technological shift which began the Bronze Age about 3500 BC. The Bronze Age was period in human cultural development when the most advanced metalworking (at least in systematic and widespread use) included techniques for smelting copper and tin from naturally occurring outcroppings of copper ores, and then smelting those ores to cast bronze. These naturally occurring ores typically included arsenic as a common impurity. Copper/tin ores are rare, as reflected in the fact that there were no tin bronzes in western Asia before 3000 BC.

After the Bronze Age, the history of metallurgy was marked by armies seeking better weaponry. Countries in Eurasia prospered when they made the superior alloys, which, in turn, made better armor and better weapons. This often determined the outcomes of battles. Significant progress in metallurgy and alchemy was made in ancient India.

Iron Age

The extraction of iron from its ore into a workable metal is much more difficult than copper or tin. It appears to have been invented by the Hittites in about 1200 BC, beginning the Iron Age. The secret of extracting and working iron was a key factor in the success of the Philistines.

In other words, the Iron Age refers to the advent of ferrous metallurgy. Historical developments in ferrous metallurgy can be found in a wide variety of past cultures and civilizations. This includes the ancient and medieval kingdoms and empires of the Middle East and Near East, ancient Iran, ancient Egypt, ancient Nubia, and Anatolia (Turkey), Ancient Nok, Carthage, the Greeks and Romans of ancient Europe, medieval Europe, ancient and medieval China, ancient and medieval India, ancient and medieval Japan, amongst others. Many applications, practices, and devices associated or involved in metallurgy were established in ancient China, such as the innovation of the blast furnace, cast iron, hydraulic-powered trip hammers, and double acting piston bellows.

Classical Antiquity and Atomism

DEMOCRITUS

Democritus, Greek philosopher of atomistic school.

Philosophical attempts to rationalize why different substances have different properties (color, density, smell), exist in different states (gaseous, liquid, and solid), and react in a different manner when exposed to environments, for example to water or fire or temperature changes, led ancient philosophers to postulate the first theories on nature and chemistry. The history of such philosophical theories that relate to chemistry can probably be traced back to every single ancient civilization. The common aspect in all these theories was the attempt to identify a small number of primary classical element that make up all the various substances in nature. Substances like air, water, and soil/earth, energy forms, such as fire and light, and more abstract concepts such as ideas, aether, and heaven, were common in ancient civilizations even in absence of any cross-fertilization; for example in Greek, Indian, Mayan, and ancient Chinese philosophies all considered air, water, earth and fire as primary elements.

Ancient World

Around 420 BC, Empedocles stated that all matter is made up of four elemental substances—earth, fire, air and water. The early theory of atomism can be traced back to ancient Greece and ancient India. Greek atomism dates back to the Greek philosopher Democritus, who declared that matter is composed of indivisible and indestructible atoms around 380 BC. Leucippus also declared that atoms were the most indivisible part of matter. This coincided with a similar declaration by Indian philosopher Kanada in his Vaisheshika sutras around the same time period. In much the same fashion he discussed the existence of gases. What Kanada declared by sutra, Democritus declared by philosophical musing. Both suffered from a lack of empirical data. Without scientific proof, the existence of atoms was easy to deny. Aristotle opposed the existence of atoms in 330 BC. Earlier, in 380 BC, a Greek text attributed to Polybus argues that the human body is composed of four humours. Around 300 BC, Epicurus postulated a universe of indestructible atoms in which man himself is responsible for achieving a balanced life.

With the goal of explaining Epicurean philosophy to a Roman audience, the Roman poet and philosopher Lucretius wrote *De Rerum Natura* (The Nature of Things) in 50 BC. In the work, Lucretius presents the principles of atomism; the nature of the mind and soul; explanations of sensation and thought; the development of the world and its phenomena; and explains a variety of celestial and terrestrial phenomena.

Much of the early development of purification methods is described by Pliny the Elder in his Naturalis Historia. He made attempts to explain those methods, as well as making acute observations of the state of many minerals.

Medieval Alchemy

The elemental system used in Medieval alchemy was developed primarily by the Arabian alchemist Jābir ibn Hayyān and rooted in the classical elements of Greek tradition. His system consisted of the four Aristotelian elements of air, earth, fire, and water in addition to two philosophical elements: sulphur, characterizing the principle of combustibility; "the stone which burns", and mercury, characterizing the principle of metallic properties. They were seen by early alchemists as idealized expressions of irreducibile components of the universe and are of larger consideration within philosophical alchemy.

Seventeenth century alchemical emblem showing the four Classical elements in the corners of the image, alongside the tria prima on the central triangle.

The three metallic principles: sulphur to flammability or combustion, mercury to volatility and stability, and salt to solidity. became the *tria prima* of the Swiss alchemist Paracelsus. He reasoned that Aristotle's four-element theory appeared in bodies as three principles. Paracelsus saw these principles as fundamental and justified them by recourse to the description of how wood burns in fire. Mercury included the cohesive principle, so that when it left in smoke the wood fell apart. Smoke described the volatility (the mercurial principle), the heat-giving flames described flammability (sulphur), and the remnant ash described solidity (salt).

The Philosopher's Stone

"The Alchemist", by Sir William Douglas, 1855

Alchemy is defined by the Hermetic quest for the philosopher's stone, the study of which is steeped in symbolic mysticism, and differs greatly from modern science. Alchemists toiled to make transformations on an esoteric (spiritual) and/or exoteric (practical) level. It was the protoscientific, exoteric aspects of alchemy that contributed heavily to the evolution of chemistry in Greco-Roman Egypt, the Islamic Golden Age, and then in Europe. Alchemy and chemistry share an interest in the composition and properties of matter, and prior to the eighteenth century were not separated into distinct disciplines. The term *chymistry* has been used to describe the blend of alchemy and chemistry that existed before this time.

The earliest Western alchemists, who lived in the first centuries of the common era, invented chemical apparatus. The *bain-marie*, or water bath is named for Mary the Jewess. Her work also gives the first descriptions of the tribikos and kerotakis. Cleopatra the Alchemist described furnaces and has been credited with the invention of the alembic. Later, the experimental framework established by Jabir ibn Hayyan influenced alchemists as the discipline migrated through the Islamic world, then to Europe in the twelfth century.

During the Renaissance, exoteric alchemy remained popular in the form of Paracelsian iatrochemistry, while spiritual alchemy flourished, realigned to its Platonic, Hermetic, and Gnostic roots. Consequently, the symbolic quest for the philosopher's stone was not superseded by scientific advances, and was still the domain of respected scientists and doctors until the early eighteenth century. Early modern alchemists who are renowned for their scientific contributions include Jan Baptist van Helmont, Robert Boyle, and Isaac Newton.

Problems Encountered with Alchemy

There were several problems with alchemy, as seen from today's standpoint. There was no systematic naming scheme for new compounds, and the language was esoteric and vague to the point that the terminologies meant different things to different people. In fact, according to *The Fontana History of Chemistry* (Brock, 1992):

The language of alchemy soon developed an arcane and secretive technical vocabulary designed to conceal information from the uninitiated. To a large degree, this language is incomprehensible to us today, though it is apparent that readers of Geoffery Chaucer's Canon's Yeoman's Tale or audiences of Ben Jonson's The Alchemist were able to construe it sufficiently to laugh at it.

Chaucer's tale exposed the more fraudulent side of alchemy, especially the manufacture of counterfeit gold from cheap substances. Less than a century earlier, Dante Alighieri also demonstrated an awareness of this fraudulence, causing him to consign all alchemists to the Inferno in his writings. Soon after, in 1317, the Avignon Pope John XXII ordered all alchemists to leave France for making counterfeit money. A law was passed in England in 1403 which made the "multiplication of metals" punishable by death. Despite these and other apparently extreme measures, alchemy did not die. Royalty and privileged classes still sought to discover the philosopher's stone and the elixir of life for themselves.

There was also no agreed-upon scientific method for making experiments reproducible. Indeed, many alchemists included in their methods irrelevant information such as the timing of the tides or the phases of the moon. The esoteric nature and codified vocabulary of alchemy appeared to be more useful in concealing the fact that they could not be sure of very much at all. As early as the

14th century, cracks seemed to grow in the facade of alchemy; and people became sceptical. Clearly, there needed to be a scientific method where experiments can be repeated by other people, and results needed to be reported in a clear language that laid out both what is known and unknown.

Alchemy in the Islamic World

In the Islamic World, the Muslims were translating the works of the ancient Greeks and Egyptians into Arabic and were experimenting with scientific ideas. The development of the modern scientific method was slow and arduous, but an early scientific method for chemistry began emerging among early Muslim chemists, beginning with the 9th century chemist Jābir ibn Hayyān (known as "Geber" in Europe), who is considered as "the father of chemistry". He introduced a systematic and experimental approach to scientific research based in the laboratory, in contrast to the ancient Greek and Egyptian alchemists whose works were largely allegorical and often unintelligble. He also invented and named the alembic (al-anbiq), chemically analyzed many chemical substances, composed lapidaries, distinguished between alkalis and acids, and manufactured hundreds of drugs. He also refined the theory of five classical elements into the theory of seven alchemical elements after identifying mercury and sulfur as chemical elements.

Jābir ibn Hayyān (Geber), a Arabian alchemist whose experimental research laid the foundations of chemistry.

Among other influential Muslim chemists, Abū al-Rayhān al-Bīrūnī, Avicenna and Al-Kindi refuted the theories of alchemy, particularly the theory of the transmutation of metals; and al-Tusi described a version of the conservation of mass, noting that a body of matter is able to change but is not able to disappear. Rhazes refuted Aristotle's theory of four classical elements for the first time and set up the firm foundations of modern chemistry, using the laboratory in the modern sense, designing and describing more than twenty instruments, many parts of which are still in use today, such as a crucible, cucurbit or retort for distillation, and the head of a still with a delivery tube (ambiq, Latin alembic), and various types of furnace or stove.

For practitioners in Europe, alchemy became an intellectual pursuit after early Arabic alchemy became available through Latin translation, and over time, they improved. Paracelsus (1493–1541), for example, rejected the 4-elemental theory and with only a vague understanding of his chemicals and medicines, formed a hybrid of alchemy and science in what was to be called iatrochemistry. Paracelsus was not perfect in making his experiments truly scientific. For example, as an extension of his theory that new compounds could be made by combining mercury with sulfur, he once made what he thought was "oil of sulfur". This was actually dimethyl ether, which had neither mercury nor sulfur.

17th and 18th centuries: Early chemistry

Agricola, author of *De re metallica*

Workroom, from *De re metallica*, 1556, Chemical Heritage Foundation

Practical attempts to improve the refining of ores and their extraction to smelt metals was an important source of information for early chemists in the 16th century, among them Georg Agricola (1494–1555), who published his great work *De re metallica* in 1556. His work describes the highly developed and complex processes of mining metal ores, metal extraction and metallurgy of the time. His approach removed the mysticism associated with the subject, creating the practical base upon which others could build. The work describes the many kinds of furnace used to smelt ore, and stimulated interest in minerals and their composition. It is no coincidence that he gives numerous references to the earlier author, Pliny the Elder and his *Naturalis Historia*. Agricola has been described as the "father of metallurgy".

In 1605, Sir Francis Bacon published *The Proficience and Advancement of Learning*, which contains a description of what would later be known as the scientific method. In 1605, Michal Sedziwój publishes the alchemical treatise *A New Light of Alchemy* which proposed the existence of the "food of life" within air, much later recognized as oxygen. In 1615 Jean Beguin published the *Tyrocinium Chymicum*, an early chemistry textbook, and in it draws the first-ever chemical equation. In 1637 René Descartes publishes *Discours de la méthode*, which contains an outline of the scientific method.

The Dutch chemist Jan Baptist van Helmont's work *Ortus medicinae* was published posthumously in 1648; the book is cited by some as a major transitional work between alchemy and chemistry, and as an important influence on Robert Boyle. The book contains the results of numerous experiments and establishes an early version of the law of conservation of mass. Working during the time just after Paracelsus and iatrochemistry, Jan Baptist van Helmont suggested that there are insubstantial substances other than air and coined a name for them - "gas", from the Greek word *chaos*. In addition to introducing the word "gas" into the vocabulary of scientists, van Helmont conducted several experiments involving gases. Jan Baptist van Helmont is also remembered today largely for his ideas on spontaneous generation and his 5-year tree experiment, as well as being considered the founder of pneumatic chemistry.

Robert Boyle

Robert Boyle, one of the co-founders of modern chemistry through his use of proper experimentation, which further separated chemistry from alchemy

Anglo-Irish chemist Robert Boyle (1627–1691) is considered to have refined the modern scientific method for alchemy and to have separated chemistry further from alchemy. Although his research clearly has its roots in the alchemical tradition, Boyle is largely regarded today as the first modern chemist, and therefore one of the founders of modern chemistry, and one of the pioneers of modern experimental scientific method. Although Boyle was not the original discover, he is best known for Boyle's law, which he presented in 1662: the law describes the inversely proportional relationship between the absolute pressure and volume of a gas, if the temperature is kept constant within a closed system.

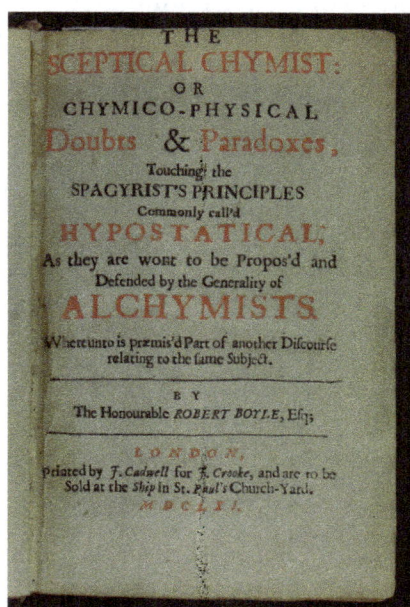

Title page from *The sceptical chymist*, 1661, Chemical Heritage Foundation

Boyle is also credited for his landmark publication *The Sceptical Chymist* in 1661, which is seen as a cornerstone book in the field of chemistry. In the work, Boyle presents his hypothesis that every phenomenon was the result of collisions of particles in motion. Boyle appealed to chemists to experiment and asserted that experiments denied the limiting of chemical elements to only the classic four: earth, fire, air, and water. He also pleaded that chemistry should cease to be subservient to medicine or to alchemy, and rise to the status of a science. Importantly, he advocated a rigorous approach to scientific experiment: he believed all theories must be proved experimentally before being regarded as true. The work contains some of the earliest modern ideas of atoms, molecules, and chemical reaction, and marks the beginning of the history of modern chemistry.

Boyle also tried to purify chemicals to obtain reproducible reactions. He was a vocal proponent of the mechanical philosophy proposed by René Descartes to explain and quantify the physical properties and interactions of material substances. Boyle was an atomist, but favoured the word *corpuscle* over *atoms*. He commented that the finest division of matter where the properties are retained is at the level of corpuscles. He also performed numerous investigations with an air pump, and noted that the mercury fell as air was pumped out. He also observed that pumping the air out of a container would extinguish a flame and kill small animals placed inside. Boyle helped to lay the foundations for the Chemical Revolution with his mechanical corpuscular philosophy. Boyle repeated the tree experiment of van Helmont, and was the first to use indicators which changed colors with acidity.

Development and Dismantling of Phlogiston

In 1702, German chemist Georg Stahl coined the name "phlogiston" for the substance believed to be released in the process of burning. Around 1735, Swedish chemist Georg Brandt analyzed a dark blue pigment found in copper ore. Brandt demonstrated that the pigment contained a new element, later named cobalt. In 1751, a Swedish chemist and pupil of Stahl's named Axel Fredrik Cronstedt, identified an impurity in copper ore as a separate metallic element, which he named nickel. Cronstedt is one of the founders of modern mineralogy. Cronstedt also discovered the mineral scheelite in 1751, which he named tungsten, meaning "heavy stone" in Swedish.

Joseph Priestley, co-discoverer of the element oxygen, which he called "dephlogisticated air"

In 1754, Scottish chemist Joseph Black isolated carbon dioxide, which he called "fixed air". In 1757, Louis Claude Cadet de Gassicourt, while investigating arsenic compounds, creates Cadet's fuming liquid, later discovered to be cacodyl oxide, considered to be the first synthetic organo-metallic compound. In 1758, Joseph Black formulated the concept of latent heat to explain the thermochemistry of phase changes. In 1766, English chemist Henry Cavendish isolated hydrogen, which he called "inflammable air". Cavendish discovered hydrogen as a colorless, odourless gas that burns and can form an explosive mixture with air, and published a paper on the production of water by burning inflammable air (that is, hydrogen) in dephlogisticated air (now known to be oxygen), the latter a constituent of atmospheric air (phlogiston theory).

In 1773, Swedish chemist Carl Wilhelm Scheele discovered oxygen, which he called "fire air", but did not immediately publish his achievement. In 1774, English chemist Joseph Priestley independently isolated oxygen in its gaseous state, calling it "dephlogisticated air", and published his work before Scheele. During his lifetime, Priestley's considerable scientific reputation rested on his invention of soda water, his writings on electricity, and his discovery of several "airs" (gases), the most famous being what Priestley dubbed "dephlogisticated air" (oxygen). However, Priestley's determination to defend phlogiston theory and to reject what would become the chemical revolution eventually left him isolated within the scientific community.

In 1781, Carl Wilhelm Scheele discovered that a new acid, tungstic acid, could be made from Cronstedt's scheelite (at the time named tungsten). Scheele and Torbern Bergman suggested that it might be possible to obtain a new metal by reducing this acid. In 1783, José and Fausto Elhuyar found an acid made from wolframite that was identical to tungstic acid. Later that year, in Spain, the brothers succeeded in isolating the metal now known as tungsten by reduction of this acid with charcoal, and they are credited with the discovery of the element.

Volta and the Voltaic Pile

A voltaic pile on display in the *Tempio Voltiano* (the Volta Temple) near Volta's home in Como.

Italian physicist Alessandro Volta constructed a device for accumulating a large charge by a series of inductions and groundings. He investigated the 1780s discovery "animal electricity" by Luigi Galvani, and found that the electric current was generated from the contact of dissimilar metals, and that the frog leg was only acting as a detector. Volta demonstrated in 1794 that when two metals and brine-soaked cloth or cardboard are arranged in a circuit they produce an electric current.

In 1800, Volta stacked several pairs of alternating copper (or silver) and zinc discs (electrodes) separated by cloth or cardboard soaked in brine (electrolyte) to increase the electrolyte conductivity. When the top and bottom contacts were connected by a wire, an electric current flowed through the voltaic pile and the connecting wire. Thus, Volta is credited with constructed the first electrical battery to produce electricity. Volta's method of stacking round plates of copper and zinc separated by disks of cardboard moistened with salt solution was termed a voltaic pile.

Thus, Volta is considered to be the founder of the discipline of electrochemistry. A Galvanic cell (or voltaic cell) is an electrochemical cell that derives electrical energy from spontaneous redox reaction taking place within the cell. It generally consists of two different metals connected by a salt bridge, or individual half-cells separated by a porous membrane.

Antoine-Laurent de Lavoisier

Although the archives of chemical research draw upon work from ancient Babylonia, Egypt, and especially the Arabs and Persians after Islam, modern chemistry flourished from the time of Antoine-Laurent de Lavoisier, a French chemist who is celebrated as the "father of modern chemistry". Lavoisier demonstrated with careful measurements that transmutation of water to earth was not possible, but that the sediment observed from boiling water came from the container. He burnt phosphorus and sulfur in air, and proved that the products weighed more than the original. Nevertheless, the weight gained was lost from the air. Thus, in 1789, he established the Law of Conservation of Mass, which is also called "Lavoisier's Law."

Portrait of Monsieur Lavoisier and his wife, by Jacques-Louis David

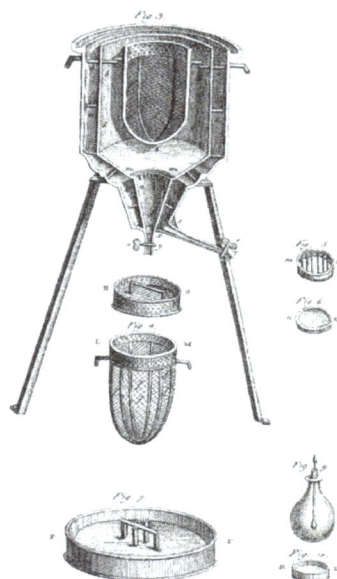

The world's first ice-calorimeter, used in the winter of 1782-83, by Antoine Lavoisier and Pierre-Simon Laplace, to determine the heat involved in various chemical changes; calculations which were based on Joseph Black's prior discovery of latent heat. These experiments mark the foundation of thermochemistry.

Repeating the experiments of Priestley, he demonstrated that air is composed of two parts, one of which combines with metals to form calxes. In *Considérations Générales sur la Nature des Acides* (1778), he demonstrated that the "air" responsible for combustion was also the source of acidity.

The next year, he named this portion oxygen (Greek for acid-former), and the other azote (Greek for no life). Lavoisier thus has a claim to the discovery of oxygen along with Priestley and Scheele. He also discovered that the "inflammable air" discovered by Cavendish - which he termed hydrogen (Greek for water-former) - combined with oxygen to produce a dew, as Priestley had reported, which appeared to be water. In *Reflexions sur le Phlogistique* (1783), Lavoisier showed the phlogiston theory of combustion to be inconsistent. Mikhail Lomonosov independently established a tradition of chemistry in Russia in the 18th century. Lomonosov also rejected the phlogiston theory, and anticipated the kinetic theory of gases. Lomonosov regarded heat as a form of motion, and stated the idea of conservation of matter.

Lavoisier worked with Claude Louis Berthollet and others to devise a system of chemical nomenclature which serves as the basis of the modern system of naming chemical compounds. In his *Methods of Chemical Nomenclature* (1787), Lavoisier invented the system of naming and classification still largely in use today, including names such as sulfuric acid, sulfates, and sulfites. In 1785, Berthollet was the first to introduce the use of chlorine gas as a commercial bleach. In the same year he first determined the elemental composition of the gas ammonia. Berthollet first produced a modern bleaching liquid in 1789 by passing chlorine gas through a solution of sodium carbonate - the result was a weak solution of sodium hypochlorite. Another strong chlorine oxidant and bleach which he investigated and was the first to produce, potassium chlorate ($KClO_3$), is known as Berthollet's Salt. Berthollet is also known for his scientific contributions to theory of chemical equilibria via the mechanism of reverse chemical reactions.

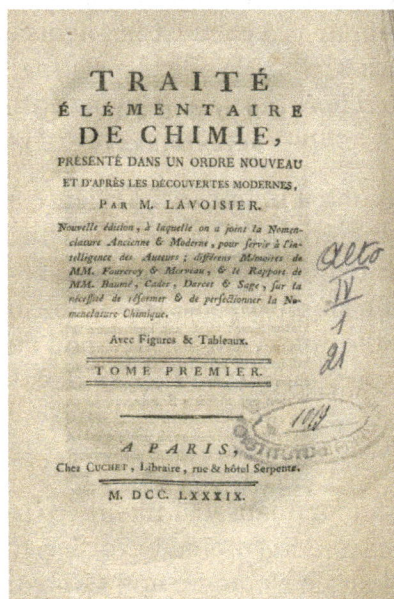

Traité élémentaire de chimie

Lavoisier's *Traité Élémentaire de Chimie* (Elementary Treatise of Chemistry, 1789) was the first modern chemical textbook, and presented a unified view of new theories of chemistry, contained a clear statement of the Law of Conservation of Mass, and denied the existence of phlogiston. In addition, it contained a list of elements, or substances that could not be broken down further, which included oxygen, nitrogen, hydrogen, phosphorus, mercury, zinc, and sulfur. His list, however, also included light, and caloric, which he believed to be material substances. In the work, Lavoisier underscored the observational basis of his chemistry, stating "I have tried...to arrive at the truth by linking up facts; to suppress as much as possible the use of reasoning, which is often an unreliable instrument which deceives us, in order to follow as much as possible the torch of observation and of experiment." Nevertheless, he believed that the real existence of atoms was philosophically impossible. Lavoisier demonstrated that organisms disassemble and reconstitute atmospheric air in the same manner as a burning body.

With Pierre-Simon Laplace, Lavoisier used a calorimeter to estimate the heat evolved per unit of carbon dioxide produced. They found the same ratio for a flame and animals, indicating that animals produced energy by a type of combustion. Lavoisier believed in the radical theory, believing that radicals, which function as a single group in a chemical reaction, would combine with oxygen in reactions. He believed all acids contained oxygen. He also discovered that diamond is a crystalline form of carbon.

While many of Lavoisier's partners were influential for the advancement of chemistry as a scientific discipline, his wife Marie-Anne Lavoisier was arguably the most influential of them all. Upon their marriage, Mme. Lavoisier began to study chemistry, English, and drawing in order to help her husband in his work either by translating papers into English, a language which Lavoisier did not know, or by keeping records and drawing the various apparatuses that Lavoisier used in his labs. Through her ability to read and translate articles from Britain for her husband, Lavoisier had access knowledge from many of the chemical advances happening outside of his lab. Furthermore, Mme. Lavoisier kept records of Lavoisier's work and ensured that his works were published. The first sign of Marie-Anne's true potential as a chemist in Lavoisier's lab came when she was translating a book by the scientist Richard Kirwan. While translating, she stumbled upon and corrected multiple errors. When she presented her translation, along with her notes to Lavoisier Her edits and contributions led to Lavoisier's refutation of the theory of phlogiston.

Lavoisier made many fundamental contributions to the science of chemistry. Following Lavoisier's work, chemistry acquired a strict quantitative nature, allowing reliable predictions to be made. The revolution in chemistry which he brought about was a result of a conscious effort to fit all experiments into the framework of a single theory. He established the consistent use of chemical balance, used oxygen to overthrow the phlogiston theory, and developed a new system of chemical nomenclature. Lavoisier was beheaded during the French Revolution.

19th Century

In 1802, French American chemist and industrialist Éleuthère Irénée du Pont, who learned manufacture of gunpowder and explosives under Antoine Lavoisier, founded a gunpowder manufacturer in Delaware known as E. I. du Pont de Nemours and Company. The French Revolution forced his family to move to the United States where du Pont started a gunpowder mill on the Brandywine River in Delaware. Wanting to make the best powder possible, du Pont was vigilant about the quality of the materials he used. For 32 years, du Pont served as president of E. I. du Pont de Nemours and Company, which eventually grew into one of the largest and most successful companies in America.

Throughout the 19th century, chemistry was divided between those who followed the atomic theory of John Dalton and those who did not, such as Wilhelm Ostwald and Ernst Mach. Although such proponents of the atomic theory as Amedeo Avogadro and Ludwig Boltzmann made great advances in explaining the behavior of gases, this dispute was not finally settled until Jean Perrin's experimental investigation of Einstein's atomic explanation of Brownian motion in the first decade of the 20th century.

Well before the dispute had been settled, many had already applied the concept of atomism to chemistry. A major example was the ion theory of Svante Arrhenius which anticipated ideas about atomic substructure that did not fully develop until the 20th century. Michael Faraday was another early worker, whose major contribution to chemistry was electrochemistry, in which (among other things) a certain quantity of electricity during electrolysis or electrodeposition of metals was shown to be associated with certain quantities of chemical elements, and fixed quantities of the elements therefore with each other, in specific ratios. These findings, like those of Dalton's combining ratios, were early clues to the atomic nature of matter.

John Dalton

John Dalton is remembered for his work on partial pressures in gases, color blindness, and atomic theory

In 1803, English meteorologist and chemist John Dalton proposed Dalton's law, which describes the relationship between the components in a mixture of gases and the relative pressure each contributes to that of the overall mixture. Discovered in 1801, this concept is also known as Dalton's law of partial pressures.

Dalton also proposed a modern atomic theory in 1803 which stated that all matter was composed of small indivisible particles termed atoms, atoms of a given element possess unique characteristics and weight, and three types of atoms exist: simple (elements), compound (simple molecules), and complex (complex molecules). In 1808, Dalton first published *New System of Chemical Philosophy* (1808-1827), in which he outlined the first modern scientific description of the atomic theory. This work identified chemical elements as a specific type of atom, therefore rejecting Newton's theory of chemical affinities.

Instead, Dalton inferred proportions of elements in compounds by taking ratios of the weights of reactants, setting the atomic weight of hydrogen to be identically one. Following Jeremias Benjamin Richter (known for introducing the term *stoichiometry*), he proposed that chemical elements combine in integral ratios. This is known as the law of multiple proportions or Dalton's law, and Dalton included a clear description of the law in his *New System of Chemical Philosophy*. The law of multiple proportions is one of the basic laws of stoichiometry used to establish the atomic theory. Despite the importance of the work as the first view of atoms as physically real entities and introduction of a system of chemical symbols, *New System of Chemical Philosophy* devoted almost as much space to the caloric theory as to atomism.

French chemist Joseph Proust proposed the law of definite proportions, which states that elements always combine in small, whole number ratios to form compounds, based on several experiments conducted between 1797 and 1804 Along with the law of multiple proportions, the law of definite proportions forms the basis of stoichiometry. The law of definite proportions and constant composition do not prove that atoms exist, but they are difficult to explain without assuming that chemical compounds are formed when atoms combine in constant proportions.

Jöns Jacob Berzelius

A Swedish chemist and disciple of Dalton, Jöns Jacob Berzelius embarked on a systematic program to try to make accurate and precise quantitative measurements and insure the purity of chemicals. Along Lavoisier, Boyle, and Dalton, Berzelius is known as the father of modern chemistry. In 1828 he compiled a table of relative atomic weights, where oxygen was set to 100, and which included all of the elements known at the time. This work provided evidence in favor of Dalton's atomic theory: that inorganic chemical compounds are composed of atoms combined in whole number amounts. He determined the exact elementary constituents of large numbers of compounds. The results strongly confirmed Proust's Law of Definite Proportions. In his weights, he used oxygen as a standard, setting its weight equal to exactly 100. He also measured the weights of 43 elements. In discovering that atomic weights are not integer multiples of the weight of hydrogen, Berzelius also disproved Prout's hypothesis that elements are built up from atoms of hydrogen.

Jöns Jacob Berzelius.

Jöns Jacob Berzelius, the chemist who worked out the modern technique of chemical formula notation and is considered one of the fathers of modern chemistry

Motivated by his extensive atomic weight determinations and in a desire to aid his experiments, he introduced the classical system of chemical symbols and notation with his 1808 publishing of *Lärbok i Kemien*, in which elements are abbreviated by one or two letters to make a distinct abbreviation from their Latin name. This system of chemical notation—in which the elements were given simple written labels, such as O for oxygen, or Fe for iron, with proportions noted by numbers—is the same basic system used today. The only difference is that instead of the subscript number used today (e.g., H_2O), Berzelius used a superscript (H^2O). Berzelius is credited with identifying the chemical elements silicon, selenium, thorium, and cerium. Students working in Berzelius's laboratory also discovered lithium and vanadium.

Berzelius developed the radical theory of chemical combination, which holds that reactions occur as stable groups of atoms called radicals are exchanged between molecules. He believed that salts are compounds of an acid and bases, and discovered that the anions in acids would be attracted to a positive electrode (the anode), whereas the cations in a base would be attracted to a negative electrode (the cathode). Berzelius did not believe in the Vitalism Theory, but instead in a regulative force which produced organization of tissues in an organism. Berzelius is also credited with originating the chemical terms "catalysis", "polymer", "isomer", and "allotrope", although his original definitions differ dramatically from modern usage. For example, he coined the term "polymer" in 1833 to describe organic compounds which shared identical empirical formulas but which differed in overall molecular weight, the larger of the compounds being described as "polymers" of the smallest. By this long superseded, pre-structural definition, glucose ($C_6H_{12}O_6$) was viewed as a polymer of formaldehyde (CH_2O).

New elements and Gas Laws

Humphry Davy, the discover of several alkali and alkaline earth metals, as well as contributions to the discoveries of the elemental nature of chlorine and iodine.

English chemist Humphry Davy was a pioneer in the field of electrolysis, using Alessandro Volta's voltaic pile to split up common compounds and thus isolate a series of new elements. He went on to electrolyse molten salts and discovered several new metals, especially sodium and potassium, highly reactive elements known as the alkali metals. Potassium, the first metal that was isolated by electrolysis, was discovered in 1807 by Davy, who derived it from caustic potash (KOH). Before the 19th century, no distinction was made between potassium and sodium. Sodium was first isolated by Davy in the same year by passing an electric current through molten sodium hydroxide (NaOH). When Davy heard that Berzelius and Pontin prepared calcium amalgam by electrolyzing lime in mercury, he tried it himself. Davy was successful, and discovered calcium in 1808 by electrolyzing a mixture of lime and mercuric oxide. He worked with electrolysis throughout his life and, in 1808, he isolated magnesium, strontium and barium.

Davy also experimented with gases by inhaling them. This experimental procedure nearly proved fatal on several occasions, but led to the discovery of the unusual effects of nitrous oxide, which came to be known as laughing gas. Chlorine was discovered in 1774 by Swedish chemist Carl Wilhelm Scheele, who called it *"dephlogisticated marine acid"* and mistakenly thought it contained oxygen. Scheele observed several properties of chlorine gas, such as its bleaching effect on litmus, its deadly effect on insects, its yellow-green colour, and the similarity of its smell to that of aqua regia. However, Scheele was unable to publish his findings at the time. In 1810, chlorine was given its current name by Humphry Davy (derived from the Greek word for green), who insisted that chlorine was in fact an element. He also showed that oxygen could not be obtained from the substance known as oxymuriatic acid (HCl solution). This discovery overturned Lavoisier's definition of acids as compounds of oxygen. Davy was a popular lecturer and able experimenter.

Joseph Louis Gay-Lussac, who stated that the ratio between the volumes of the reactant gases and the products can be expressed in simple whole numbers.

French chemist Joseph Louis Gay-Lussac shared the interest of Lavoisier and others in the quantitative study of the properties of gases. From his first major program of research in 1801–1802, he concluded that equal volumes of all gases expand equally with the same increase in temperature: this conclusion is usually called "Charles's law", as Gay-Lussac gave credit to Jacques Charles, who had arrived at nearly the same conclusion in the 1780s but had not published it. The law was independently discovered by British natural philosopher John Dalton by 1801, although Dalton's description was less thorough than Gay-Lussac's. In 1804 Gay-Lussac made several daring ascents of over 7,000 meters above sea level in hydrogen-filled balloons—a feat not equaled for another 50 years—that allowed him to investigate other aspects of gases. Not only did he gather magnetic measurements at various altitudes, but he also took pressure, temperature, and humidity measurements and samples of air, which he later analyzed chemically.

In 1808 Gay-Lussac announced what was probably his single greatest achievement: from his own and others' experiments he deduced that gases at constant temperature and pressure combine in simple numerical proportions by volume, and the resulting product or products—if gases—also bear a simple proportion by volume to the volumes of the reactants. In other words, gases under equal conditions of temperature and pressure react with one another in volume ratios of small whole numbers. This conclusion subsequently became known as "Gay-Lussac's law" or the "Law of Combining Volumes". With his fellow professor at the École Polytechnique, Louis Jacques Thénard, Gay-Lussac also participated in early electrochemical research, investigating the elements discovered by its means. Among other achievements, they decomposed boric acid by using fused potassium, thus discovering the element boron. The two also took part in contemporary debates that modified Lavoisier's definition of acids and furthered his program of analyzing organic compounds for their oxygen and hydrogen content.

The element iodine was discovered by French chemist Bernard Courtois in 1811. Courtois gave samples to his friends, Charles Bernard Desormes (1777–1862) and Nicolas Clément (1779–1841), to continue research. He also gave some of the substance to Gay-Lussac and to physicist André-Marie Ampère. On December 6, 1813, Gay-Lussac announced that the new substance was either an element or a compound of oxygen. It was Gay-Lussac who suggested the name *"iode"* for violet (because of the color of iodine vapor). Ampère had given some of his sample to Humphry Davy. Davy did some experiments on the substance and noted its similarity to chlorine. Davy sent a letter dated December 10 to the Royal Society of London stating that he had identified a new element. Arguments erupted between Davy and Gay-Lussac over who identified iodine first, but both scientists acknowledged Courtois as the first to isolate the element.

Amedeo Avogadro, who postulated that, under controlled conditions of temperature and pressure, equal volumes of gases contain an equal number of molecules. This is known as Avogadro's law.

In 1815, Humphry Davy invented the Davy lamp, which allowed miners within coal mines to work safely in the presence of flammable gases. There had been many mining explosions caused by firedamp or methane often ignited by open flames of the lamps then used by miners. Davy conceived of using an iron gauze to enclose a lamp's flame, and so prevent the methane burning inside the lamp from passing out to the general atmosphere. Although the idea of the safety lamp had already been demonstrated by William Reid Clanny and by the then unknown (but later very famous) engineer George Stephenson, Davy's use of wire gauze to prevent the spread of flame was used by many other inventors in their later designs. There was some discussion as to whether Davy had discovered the principles behind his lamp without the help of the work of Smithson Tennant, but it was generally agreed that the work of both men had been independent. Davy refused to patent the lamp, and its invention led to him being awarded the Rumford medal in 1816.

After Dalton published his atomic theory in 1808, certain of his central ideas were soon adopted by most chemists. However, uncertainty persisted for half a century about how atomic theory was to be configured and applied to concrete situations; chemists in different countries developed several different incompatible atomistic systems. A paper that suggested a way out of this difficult situation was published as early as 1811 by the Italian physicist Amedeo Avogadro (1776-1856), who hypothesized that equal volumes of gases at the same temperature and pressure contain equal numbers of molecules, from which it followed that relative molecular weights of any two gases are the same as the ratio of the densities of the two gases under the same conditions of temperature and pressure. Avogadro also reasoned that simple gases were not formed of solitary atoms but were instead compound molecules of two or more atoms. Thus Avogadro was able to overcome the difficulty that Dalton and others had encountered when Gay-Lussac reported that above 100 °C the volume of water vapor was twice the volume of the oxygen used to form it. According to Avogadro, the molecule of oxygen had split into two atoms in the course of forming water vapor.

Avogadro's hypothesis was neglected for half a century after it was first published. Many reasons for this neglect have been cited, including some theoretical problems, such as Jöns Jacob Berzelius's "dualism", which asserted that compounds are held together by the attraction of positive and negative electrical charges, making it inconceivable that a molecule composed of two electrically similar atoms—as in oxygen—could exist. An additional barrier to acceptance was the fact that many chemists were reluctant to adopt physical methods (such as vapour-density determinations) to solve their problems. By mid-century, however, some leading figures had begun to view the chaotic multiplicity of competing systems of atomic weights and molecular formulas as intolerable. Moreover, purely chemical evidence began to mount that suggested Avogadro's approach might be right after all. During the 1850s, younger chemists, such as Alexander Williamson in England, Charles Gerhardt and Charles-Adolphe Wurtz in France, and August Kekulé in Germany, began to advocate reforming theoretical chemistry to make it consistent with Avogadrian theory.

Wöhler and the Vitalism Debate

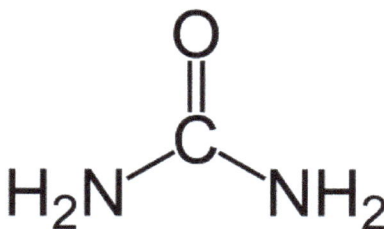

Structural formula of urea

In 1825, Friedrich Wöhler and Justus von Liebig performed the first confirmed discovery and explanation of isomers, earlier named by Berzelius. Working with cyanic acid and fulminic acid, they correctly deduced that isomerism was caused by differing arrangements of atoms within a molecular structure. In 1827, William Prout classified biomolecules into their modern groupings: carbohydrates, proteins and lipids. After the nature of combustion was settled, another dispute, about vitalism and the essential distinction between organic and inorganic substances, began. The vitalism question was revolutionized in 1828 when Friedrich Wöhler synthesized urea, thereby establishing that organic compounds could be produced from inorganic starting materials and disproving the theory of vitalism. Never before had an organic compound been synthesized from inorganic material.

This opened a new research field in chemistry, and by the end of the 19th century, scientists were able to synthesize hundreds of organic compounds. The most important among them are mauve, magenta, and other synthetic dyes, as well as the widely used drug aspirin. The discovery of the artificial synthesis of urea contributed greatly to the theory of isomerism, as the empirical chemical formulas for urea and ammonium cyanate are identical. In 1832, Friedrich Wöhler and Justus von Liebig discovered and explained functional groups and radicals in relation to organic chemistry, as well as first synthesizing benzaldehyde. Liebig, a German chemist, made major contributions to agricultural and biological chemistry, and worked on the organization of organic chemistry. Liebig is considered the "father of the fertilizer industry" for his discovery of nitrogen as an essential plant nutrient, and his formulation of the Law of the Minimum which described the effect of individual nutrients on crops.

Mid-1800s

In 1840, Germain Hess proposed Hess's law, an early statement of the law of conservation of energy, which establishes that energy changes in a chemical process depend only on the states of the starting and product materials and not on the specific pathway taken between the two states. In 1847, Hermann Kolbe obtained acetic acid from completely inorganic sources, further disproving vitalism. In 1848, William Thomson, 1st Baron Kelvin (commonly known as Lord Kelvin) established the concept of absolute zero, the temperature at which all molecular motion ceases. In 1849, Louis Pasteur discovered that the racemic form of tartaric acid is a mixture of the levorotatory and dextrotatory forms, thus clarifying the nature of optical rotation and advancing the field of stereochemistry. In 1852, August Beer proposed Beer's law, which explains the relationship between the composition of a mixture and the amount of light it will absorb. Based partly on earlier work by Pierre Bouguer and Johann Heinrich Lambert, it established the analytical technique known as spectrophotometry. In 1855, Benjamin Silliman, Jr. pioneered methods of petroleum cracking, which made the entire modern petrochemical industry possible.

Avogadro's hypothesis began to gain broad appeal among chemists only after his compatriot and fellow scientist Stanislao Cannizzaro demonstrated its value in 1858, two years after Avogadro's death. Cannizzaro's chemical interests had originally centered on natural products and on reactions of aromatic compounds; in 1853 he discovered that when benzaldehyde is treated with concentrated base, both benzoic acid and benzyl alcohol are produced—a phenomenon known today as the Cannizzaro reaction. In his 1858 pamphlet, Cannizzaro showed that a complete return to the ideas of Avogadro could be used to construct a consistent and robust theoretical structure that fit nearly all of the available empirical evidence. For instance, he pointed to evidence that suggested that not all elementary gases consist of two atoms per molecule—some were monatomic, most were diatomic, and a few were even more complex.

$C_4H_4O_4$ . . . . empirische Formel.

$C_4H_3O_3 + HO$ . . . . . . dualistische Formel.

$C_4H_3O_4 . H$ . . . . Wasserstoffsäure-Theorie.

$C_4H_4 + O_4$ . . . . . . Kerntheorie.

$C_4H_3O_2 + HO_2$ . . . . . . Longchamp's Ansicht.

$C_4H + H_3O_4$ . . . . . . Graham's Ansicht.

$C_4H_3O_2.O + HO$ . . . . . Radicaltheorie

$C_4H_3 . O_3 + HO$ . . . . . Radicaltheorie.

$\left.{C_4H_3O_2 \atop H}\right\}O_2$ . . . . Gerhardt. Typentheorie.

$\left.{C_2H_3 \atop H}\right\}O_4$ . . . . Typentheorie (Schischkoff) etc.

$C_2O_3 + C_2H_3 + HO$ . . . . Berzelius' Paarlingstheorie.

$HO.(C_2H_3)C_2, O_3$ . . . . . Kolbe's Ansicht.

$HO.(C_2H_3)C_2, O.O_2$ . . . . ditto

$\left.{C_2(C_2H_3)O_2 \atop H}\right\}O_2$ . . . . Wurtz.

$\left.{C_2H_3(C_2O_3) \atop H}\right\}O_2$ . . . . . . Mendius.

$\left.{C_2H_2.HO \atop HO}\right\}C_2O_2$ . . . . . . Geuther.

$C_2\left\{{C_2H_3 \atop O \atop O}\right\}O + HO$ . . . . Rochleder.

$\left(C_2 {H_3 \atop CO} + CO_2\right) + HO$. Persoz.

$C_2\left\{{C_2\{^{O_2}_{H^2} \atop H} \atop {H \atop \overline{H}}}\right\}O_2$ . . . . . . Buff.

Formulas of acetic acid given by August Kekulé in 1861.

Another point of contention had been the formulas for compounds of the alkali metals (such as sodium) and the alkaline earth metals (such as calcium), which, in view of their striking chemical analogies, most chemists had wanted to assign to the same formula type. Cannizzaro argued that placing these metals in different categories had the beneficial result of eliminating certain anomalies when using their physical properties to deduce atomic weights. Unfortunately, Cannizzaro's pamphlet was published initially only in Italian and had little immediate impact. The real breakthrough came with an international chemical congress held in the German town of Karlsruhe in September 1860, at which most of the leading European chemists were present. The Karlsruhe Congress had been arranged by Kekulé, Wurtz, and a few others who shared Cannizzaro's sense of the direction chemistry should go. Speaking in French (as everyone there did), Cannizzaro's eloquence and logic made an indelible impression on the assembled body. Moreover, his friend Angelo Pavesi distributed Cannizzaro's pamphlet to attendees at the end of the meeting; more than one chemist later wrote of the decisive impression the reading of this document provided. For instance, Lothar Meyer later wrote that on reading Cannizzaro's paper, "The scales seemed to fall from my eyes." Cannizzaro thus played a crucial role in winning the battle for reform. The system advocated by him, and soon thereafter adopted by most leading chemists, is substantially identical to what is still used today.

Perkin, Crookes, and Nobel

In 1856, Sir William Henry Perkin, age 18, given a challenge by his professor, August Wilhelm von Hofmann, sought to synthesize quinine, the anti-malaria drug, from coal tar. In one attempt, Perkin oxidized aniline using potassium dichromate, whose toluidine impurities reacted with the aniline and yielded a black solid—suggesting a "failed" organic synthesis. Cleaning the flask with alcohol, Perkin noticed purple portions of the solution: a byproduct of the attempt was the first synthetic dye, known as mauveine or Perkin's mauve. Perkin's discovery is the foundation of the dye synthesis industry, one of the earliest successful chemical industries.

disorder in the field for his students. Mendeleev found that, when all the known chemical elements were arranged in order of increasing atomic weight, the resulting table displayed a recurring pattern, or periodicity, of properties within groups of elements. Mendeleev's law allowed him to build up a systematic periodic table of all the 66 elements then known based on atomic mass, which he published in *Principles of Chemistry* in 1869. His first Periodic Table was compiled on the basis of arranging the elements in ascending order of atomic weight and grouping them by similarity of properties.

Mendeleev had such faith in the validity of the periodic law that he proposed changes to the generally accepted values for the atomic weight of a few elements and, in his version of the periodic table of 1871, predicted the locations within the table of unknown elements together with their properties. He even predicted the likely properties of three yet-to-be-discovered elements, which he called ekaboron (Eb), ekaaluminium (Ea), and ekasilicon (Es), which proved to be good predictors of the properties of scandium, gallium, and germanium, respectively, which each fill the spot in the periodic table assigned by Mendeleev.

At first the periodic system did not raise interest among chemists. However, with the discovery of the predicted elements, notably gallium in 1875, scandium in 1879, and germanium in 1886, it began to win wide acceptance. The subsequent proof of many of his predictions within his lifetime brought fame to Mendeleev as the founder of the periodic law. This organization surpassed earlier attempts at classification by Alexandre-Émile Béguyer de Chancourtois, who published the telluric helix, an early, three-dimensional version of the periodic table of the elements in 1862, John Newlands, who proposed the law of octaves (a precursor to the periodic law) in 1864, and Lothar Meyer, who developed an early version of the periodic table with 28 elements organized by valence in 1864. Mendeleev's table did not include any of the noble gases, however, which had not yet been discovered. Gradually the periodic law and table became the framework for a great part of chemical theory. By the time Mendeleev died in 1907, he enjoyed international recognition and had received distinctions and awards from many countries.

In 1873, Jacobus Henricus van 't Hoff and Joseph Achille Le Bel, working independently, developed a model of chemical bonding that explained the chirality experiments of Pasteur and provided a physical cause for optical activity in chiral compounds. van 't Hoff's publication, called *Voorstel tot Uitbreiding der Tegenwoordige in de Scheikunde gebruikte Structuurformules in de Ruimte*, etc. (Proposal for the development of 3-dimensional chemical structural formulae) and consisting of twelve pages text and one page diagrams, gave the impetus to the development of stereochemistry. The concept of the "asymmetrical carbon atom", dealt with in this publication, supplied an explanation of the occurrence of numerous isomers, inexplicable by means of the then current structural formulae. At the same time he pointed out the existence of relationship between optical activity and the presence of an asymmetrical carbon atom.

Josiah Willard Gibbs

American mathematical physicist J. Willard Gibbs's work on the applications of thermodynamics was instrumental in transforming physical chemistry into a rigorous deductive science. During the years from 1876 to 1878, Gibbs worked on the principles of thermodynamics, applying them to the complex processes involved in chemical reactions. He discovered the concept of chemical potential, or the "fuel" that makes chemical reactions work. In 1876 he published his most famous contribution, "On the Equilibrium of Heterogeneous Substances", a compilation of his work on thermodynamics and physical chemistry which laid out the concept of free energy to explain the physical basis of chemical equilibria. In these essays were the beginnings of Gibbs' theories of phases of matter: he considered each state of matter a phase, and each substance a component. Gibbs took all of the variables involved in a chemical reaction - temperature, pressure, energy, volume, and entropy - and included them in one simple equation known as Gibbs' phase rule.

J. Willard Gibbs formulated a concept of thermodynamic equilibrium of a system in terms of energy and entropy. He also did extensive work on chemical equilibrium, and equilibria between phases.

Within this paper was perhaps his most outstanding contribution, the introduction of the concept free energy, now universally called Gibbs free energy in his honor. The Gibbs free energy relates the tendency of a physical or chemical system to simultaneously lower its energy and increase its disorder, or entropy, in a spontaneous natural process. Gibbs's approach allows a researcher to calculate the change in free energy in the process, such as in a chemical reaction, and how fast it will happen. Since virtually all chemical processes and many physical ones involve such changes, his work has significantly impacted both the theoretical and experiential aspects of these sciences. In 1877, Ludwig Boltzmann established statistical derivations of many important physical and chemical concepts, including entropy, and distributions of molecular velocities in the gas phase. Together with Boltzmann and James Clerk Maxwell, Gibbs created a new branch of theoretical physics called statistical mechanics (a term that he coined), explaining the laws of thermodynamics as consequences of the statistical properties of large ensembles of particles. Gibbs also worked on the application of Maxwell's equations to problems in physical optics. Gibbs's derivation of the phenomenological laws of thermodynamics from the statistical properties of systems with many particles was presented in his highly influential textbook *Elementary Principles in Statistical Mechanics*, published in 1902, a year before his death. In that work, Gibbs reviewed the relationship between the laws of thermodynamics and statistical theory of molecular motions. The overshooting of the original function by partial sums of Fourier series at points of discontinuity is known as the Gibbs phenomenon.

Late 19th Century

German engineer Carl von Linde's invention of a continuous process of liquefying gases in large quantities formed a basis for the modern technology of refrigeration and provided both impetus and means for conducting scientific research at low temperatures and very high vacuums. He developed a methyl ether refrigerator (1874) and an ammonia refrigerator (1876). Though other refrigeration units had been developed earlier, Linde's were the first to be designed with the aim of precise calculations of efficiency. In 1895 he set up a large-scale plant for the production of liquid air. Six years later he developed a method for separating pure liquid oxygen from liquid air that resulted in widespread industrial conversion to processes utilizing oxygen (e.g., in steel manufacture).

In 1883, Svante Arrhenius developed an ion theory to explain conductivity in electrolytes. In 1884, Jacobus Henricus van 't Hoff published *Études de Dynamique chimique* (Studies in Dynamic Chemistry), a seminal study on chemical kinetics. In this work, van 't Hoff entered for the first time the field of physical chemistry. Of great importance was his development of the general thermodynamic relationship between the heat of conversion and the displacement of the equilibrium as a result of temperature variation. At constant volume, the equilibrium in a system will tend to shift in such a direction as to oppose the temperature change which is imposed upon the system. Thus, lowering the temperature results in heat development while increasing the temperature results in heat absorption. This principle of mobile equilibrium was subsequently (1885) put in a general form by Henry Louis Le Chatelier, who extended the principle to include compensation, by change of volume, for imposed pressure changes. The van 't Hoff-Le Chatelier principle, or simply Le Chatelier's principle, explains the response of dynamic chemical equilibria to external stresses.

In 1884, Hermann Emil Fischer proposed the structure of purine, a key structure in many biomolecules, which he later synthesized in 1898. He also began work on the chemistry of glucose and related sugars. In 1885, Eugene Goldstein named the cathode ray, later discovered to be composed of electrons, and the canal ray, later discovered to be positive hydrogen ions that had been stripped of their electrons in a cathode ray tube; these would later be named protons. The year 1885 also saw the publishing of J. H. van 't Hoff's *L'Équilibre chimique dans les Systèmes gazeux ou dissous à l'État dilué* (Chemical equilibria in gaseous systems or strongly diluted solutions), which dealt with this theory of dilute solutions. Here he demonstrated that the "osmotic pressure" in solutions which are sufficiently dilute is proportionate to the concentration and the absolute temperature so that this pressure can be represented by a formula which only deviates from the formula for gas pressure by a coefficient **i**. He also determined the value of *i* by various methods, for example by means of the vapor pressure and François-Marie Raoult's results on the lowering of the freezing point. Thus van 't Hoff was able to prove that thermodynamic laws are not only valid for gases, but also for dilute solutions. His pressure laws, given general validity by the electrolytic dissociation theory of Arrhenius (1884-1887) - the first foreigner who came to work with him in Amsterdam (1888) - are considered the most comprehensive and important in the realm of natural sciences. In 1893, Alfred Werner discovered the octahedral structure of cobalt complexes, thus establishing the field of coordination chemistry.

Ramsay's Discovery of the Noble Gases

The most celebrated discoveries of Scottish chemist William Ramsay were made in inorganic chemistry. Ramsay was intrigued by the British physicist John Strutt, 3rd Baron Rayleigh's 1892 discovery that the atomic weight of nitrogen found in chemical compounds was lower than that of nitrogen found in the atmosphere. He ascribed this discrepancy to a light gas included in chemical compounds of nitrogen, while Ramsay suspected a hitherto undiscovered heavy gas in atmospheric nitrogen. Using two different methods to remove all known gases from air, Ramsay and Lord Rayleigh were able to announce in 1894 that they had found a monatomic, chemically inert gaseous element that constituted nearly 1 percent of the atmosphere; they named it argon.

The following year, Ramsay liberated another inert gas from a mineral called cleveite; this proved to be helium, previously known only in the solar spectrum. In his book *The Gases of the Atmosphere* (1896), Ramsay showed that the positions of helium and argon in the periodic table of elements indicated that at least three more noble gases might exist. In 1898 Ramsay and the British chemist Morris W. Travers isolated these elements—called neon, krypton, and xenon—from air brought to a liquid state at low temperature and high pressure. Sir William Ramsay worked with Frederick Soddy to demonstrate, in 1903, that alpha particles (helium nuclei) were continually produced during the radioactive decay of a sample of radium. Ramsay was awarded the 1904 Nobel Prize for Chemistry in recognition of "services in the discovery of the inert gaseous elements in air, and his determination of their place in the periodic system."

In 1897, J. J. Thomson discovered the electron using the cathode ray tube. In 1898, Wilhelm Wien demonstrated that canal rays (streams of positive ions) can be deflected by magnetic fields, and that the amount of deflection is proportional to the mass-to-charge ratio. This discovery would lead to the analytical technique known as mass spectrometry.

Marie and Pierre Curie

Marie Curie, a pioneer in the field of radioactivity and the first twice-honored Nobel laureate (and still the only one in two different sciences)

Pierre Curie, known for his work on radioactivity as well as on ferromagnetism, paramagnetism, and diamagnetism; notably Curie's law and Curie point.

Marie Skłodowska-Curie was a Polish-born French physicist and chemist who is famous for her pioneering research on radioactivity. She and her husband are considered to have laid the cornerstone of the nuclear age with their research on radioactivity. Marie was fascinated with the work of Henri Becquerel, a French physicist who discovered in 1896 that uranium casts off rays similar to the X-rays discovered by Wilhelm Röntgen. Marie Curie began studying uranium in late 1897 and theorized, according to a 1904 article she wrote for Century magazine, "that the emission of rays by the compounds of uranium is a property of the metal itself—that it is an atomic property of the element uranium independent of its chemical or physical state." Curie took Becquerel's work a few steps further, conducting her own experiments on uranium rays. She discovered that the rays remained constant, no matter the condition or form of the uranium. The rays, she theorized, came from the element's atomic structure. This revolutionary idea created the field of atomic physics and the Curies coined the word *radioactivity* to describe the phenomena.

Pierre and Marie further explored radioactivity by working to separate the substances in uranium ores and then using the electrometer to make radiation measurements to 'trace' the minute amount of unknown radioactive element among the fractions that resulted. Working with the mineral pitchblende, the pair discovered a new radioactive element in 1898. They named the element polonium, after Marie's native country of Poland. On December 21, 1898, the Curies detected the presence of another radioactive material in the pitchblende. They presented this finding to the French Academy of Sciences on December 26, proposing that the new element be called radium. The Curies then went to work isolating polonium and radium from naturally occurring compounds to prove that they were new elements. In 1902, the Curies announced that they had produced a decigram of pure radium, demonstrating its existence as a unique chemical element. While it took three years for them to isolate radium, they were never able to isolate polonium. Along with the discovery of two new elements and finding techniques for isolating radioactive isotopes, Curie oversaw the world's first studies into the treatment of neoplasms, using radioactive isotopes. With Henri Becquerel and her husband, Pierre Curie, she was awarded the 1903 Nobel Prize for Physics. She was the sole winner of the 1911 Nobel Prize for Chemistry. She was the first woman to win a Nobel Prize, and she is the only woman to win the award in two different fields.

While working with Marie to extract pure substances from ores, an undertaking that really required industrial resources but that they achieved in relatively primitive conditions, Pierre himself concentrated on the physical study (including luminous and chemical effects) of the new radiations. Through the action of magnetic fields on the rays given out by the radium, he proved the existence of particles electrically positive, negative, and neutral; these Ernest Rutherford was afterward to call alpha, beta, and gamma rays. Pierre then studied these radiations by calorimetry and also observed the physiological effects of radium, thus opening the way to radium therapy. Among Pierre Curie's discoveries were that ferromagnetic substances exhibited a critical temperature transition, above which the substances lost their ferromagnetic behavior - this is known as the "Curie point." He was elected to the Academy of Sciences (1905), having in 1903 jointly with Marie received the Royal Society's prestigious Davy Medal and jointly with her and Becquerel the Nobel Prize for Physics. He was run over by a carriage in the rue Dauphine in Paris in 1906 and died instantly. His complete works were published in 1908.

Ernest Rutherford

New Zealand-born chemist and physicist Ernest Rutherford is considered to be "the father of nuclear physics." Rutherford is best known for devising the names alpha, beta, and gamma to classify various forms of radioactive "rays" which were poorly understood at his time (alpha and beta rays are particle beams, while gamma rays are a form of high-energy electromagnetic radiation). Rutherford deflected alpha rays with both electric and magnetic fields in 1903. Working with Frederick Soddy, Rutherford explained that radioactivity is due to the transmutation of elements, now known to involve nuclear reactions.

Ernest Rutherford, discoverer of the nucleus and considered the father of nuclear physics

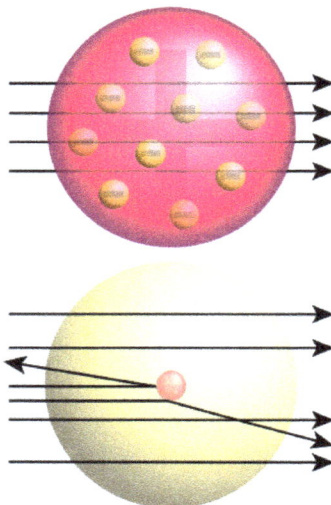

Top: Predicted results based on the then-accepted plum pudding model of the atom. Bottom: Observed results. Rutherford disproved the plum pudding model and concluded that the positive charge of the atom must be concentrated in a small, central nucleus.

He also observed that the intensity of radioactivity of a radioactive element decreases over a unique and regular amount of time until a point of stability, and he named the halving time the "half-life." In 1901 and 1902 he worked with Frederick Soddy to prove that atoms of one radioactive element would spontaneously turn into another, by expelling a piece of the atom at high velocity. In 1906 at the University of Manchester, Rutherford oversaw an experiment conducted by his students Hans Geiger (known for the Geiger counter) and Ernest Marsden. In the Geiger–Marsden experiment, a beam of alpha particles, generated by the radioactive decay of radon, was directed normally onto a sheet of very thin gold foil in an evacuated chamber. Under the prevailing plum pudding model, the alpha particles should all have passed through the foil and hit the detector screen, or have been deflected by, at most, a few degrees.

However, the actual results surprised Rutherford. Although many of the alpha particles did pass through as expected, many others were deflected at small angles while others were reflected back to the alpha source. They observed that a very small percentage of particles were deflected through angles much larger than 90 degrees. The gold foil experiment showed large deflections for a small fraction of incident particles. Rutherford realized that, because some of the alpha particles were deflected or reflected, the atom had a concentrated centre of positive charge and of relatively large mass - Rutherford later termed this positive center the "atomic nucleus". The alpha particles had either hit the positive centre directly or passed by it close enough to be affected by its positive charge. Since many other particles passed through the gold foil, the positive centre would have to be a relatively small size compared to the rest of the atom - meaning that the atom is mostly open space. From his results, Rutherford developed a model of the atom that was similar to the solar system, known as Rutherford model. Like planets, electrons orbited a central, sun-like nucleus. For his work with radiation and the atomic nucleus, Rutherford received the 1908 Nobel Prize in Chemistry.

20th Century

The first Solvay Conference was held in Brussels in 1911 and was considered a turning point in the world of physics and chemistry.

In 1903, Mikhail Tsvet invented chromatography, an important analytic technique. In 1904, Hantaro Nagaoka proposed an early nuclear model of the atom, where electrons orbit a dense massive nucleus. In 1905, Fritz Haber and Carl Bosch developed the Haber process for making ammonia, a milestone in industrial chemistry with deep consequences in agriculture. The Haber process, or Haber-Bosch process, combined nitrogen and hydrogen to form ammonia in industrial quantities for production of fertilizer and munitions. The food production for half the world's current population depends on this method for producing fertilizer. Haber, along with Max Born, proposed the Born–Haber cycle as a method for evaluating the lattice energy of an ionic solid. Haber has also been described as the "father of chemical warfare" for his work developing and deploying chlorine and other poisonous gases during World War I.

In 1905, Albert Einstein explained Brownian motion in a way that definitively proved atomic theory. Leo Baekeland invented bakelite, one of the first commercially successful plastics. In 1909, American physicist Robert Andrews Millikan - who had studied in Europe under Walther Nernst and Max Planck - measured the charge of individual electrons with unprecedented accuracy through the oil drop experiment, in which he measured the electric charges on tiny falling water (and later oil) droplets. His study established that any particular droplet's electrical charge is a multiple of a definite, fundamental value — the electron's charge — and thus a confirmation that all electrons have the same charge and mass. Beginning in 1912, he spent several years investigating and finally proving Albert Einstein's proposed linear relationship between energy and frequency, and providing the first direct photoelectric support for Planck's constant. In 1923 Millikan was awarded the Nobel Prize for Physics.

Robert A. Millikan, who is best known for measuring the charge on the electron,
won the Nobel Prize in Physics in 1923.

In 1909, S. P. L. Sørensen invented the pH concept and develops methods for measuring acidity. In 1911, Antonius Van den Broek proposed the idea that the elements on the periodic table are more properly organized by positive nuclear charge rather than atomic weight. In 1911, the first Solvay Conference was held in Brussels, bringing together most of the most prominent scientists of the day. In 1912, William Henry Bragg and William Lawrence Bragg proposed Bragg's law and established the field of X-ray crystallography, an important tool for elucidating the crystal structure of substances. In 1912, Peter Debye develops the concept of molecular dipole to describe asymmetric charge distribution in some molecules.

Niels Bohr

In 1913, Niels Bohr, a Danish physicist, introduced the concepts of quantum mechanics to atomic structure by proposing what is now known as the Bohr model of the atom, where electrons exist only in strictly defined circular orbits around the nucleus similar to rungs on a ladder. The Bohr Model is a planetary model in which the negatively charged electrons orbit a small, positively charged nucleus similar to the planets orbiting the Sun (except that the orbits are not planar) - the gravitational force of the solar system is mathematically akin to the attractive Coulomb (electrical) force between the positively charged nucleus and the negatively charged electrons.

In the Bohr model, however, electrons orbit the nucleus in orbits that have a set size and energy -

the energy levels are said to be *quantized*, which means that only certain orbits with certain radii are allowed; orbits in between simply don't exist. The energy of the orbit is related to its size - that is, the lowest energy is found in the smallest orbit. Bohr also postulated that electromagnetic radiation is absorbed or emitted when an electron moves from one orbit to another. Because only certain electron orbits are permitted, the emission of light accompanying a jump of an electron from an excited energy state to ground state produces a unique emission spectrum for each element.

Niels Bohr, the developer of the Bohr model of the atom, and a leading founder of quantum mechanics

Niels Bohr also worked on the principle of complementarity, which states that an electron can be interpreted in two mutually exclusive and valid ways. Electrons can be interpreted as wave or particle models. His hypothesis was that an incoming particle would strike the nucleus and create an excited compound nucleus. This formed the basis of his liquid drop model and later provided a theory base for the explanation of nuclear fission.

In 1913, Henry Moseley, working from Van den Broek's earlier idea, introduces concept of atomic number to fix inadequacies of Mendeleev's periodic table, which had been based on atomic weight. The peak of Frederick Soddy's career in radiochemistry was in 1913 with his formulation of the concept of isotopes, which stated that certain elements exist in two or more forms which have different atomic weights but which are indistinguishable chemically. He is remembered for proving the existence of isotopes of certain radioactive elements, and is also credited, along with others, with the discovery of the element protactinium in 1917. In 1913, J. J. Thomson expanded on the work of Wien by showing that charged subatomic particles can be separated by their mass-to-charge ratio, a technique known as mass spectrometry.

Gilbert N. Lewis

American physical chemist Gilbert N. Lewis laid the foundation of valence bond theory; he was instrumental in developing a bonding theory based on the number of electrons in the outermost "valence" shell of the atom. In 1902, while Lewis was trying to explain valence to his students, he depicted atoms as constructed of a concentric series of cubes with electrons at each corner. This "cubic atom" explained the eight groups in the periodic table and represented his idea that chemical bonds are formed by electron transference to give each atom a complete set of eight outer electrons (an "octet").

Lewis's theory of chemical bonding continued to evolve and, in 1916, he published his seminal article "The Atom of the Molecule", which suggested that a chemical bond is a pair of electrons shared by two atoms. Lewis's model equated the classical chemical bond with the sharing of a pair of electrons between the two bonded atoms. Lewis introduced the "electron dot diagrams" in this paper to symbolize the electronic structures of atoms and molecules. Now known as Lewis structures, they are discussed in virtually every introductory chemistry book.

Shortly after publication of his 1916 paper, Lewis became involved with military research. He did not return to the subject of chemical bonding until 1923, when he masterfully summarized his model in a short monograph entitled Valence and the Structure of Atoms and Molecules. His renewal of interest in this subject was largely stimulated by the activities of the American chemist and General Electric researcher Irving Langmuir, who between 1919 and 1921 popularized and elaborated Lewis's model. Langmuir subsequently introduced the term *covalent bond*. In 1921, Otto Stern and Walther Gerlach establish concept of quantum mechanical spin in subatomic particles.

For cases where no sharing was involved, Lewis in 1923 developed the electron pair theory of acids and base: Lewis redefined an acid as any atom or molecule with an incomplete octet that was thus capable of accepting electrons from another atom; bases were, of course, electron donors. His theory is known as the concept of Lewis acids and bases. In 1923, G. N. Lewis and Merle Randall published *Thermodynamics and the Free Energy of Chemical Substances*, first modern treatise on chemical thermodynamics.

The 1920s saw a rapid adoption and application of Lewis's model of the electron-pair bond in the fields of organic and coordination chemistry. In organic chemistry, this was primarily due to the efforts of the British chemists Arthur Lapworth, Robert Robinson, Thomas Lowry, and Christopher Ingold; while in coordination chemistry, Lewis's bonding model was promoted through the efforts of the American chemist Maurice Huggins and the British chemist Nevil Sidgwick.

Quantum Mechanics

In 1924, French quantum physicist Louis de Broglie published his thesis, in which he introduced a revolutionary theory of electron waves based on wave–particle duality in his thesis. In his time, the wave and particle interpretations of light and matter were seen as being at odds with one another, but de Broglie suggested that these seemingly different characteristics were instead the same behavior observed from different perspectives — that particles can behave like waves, and waves (radiation) can behave like particles. Broglie's proposal offered an explanation of the restriction motion of electrons within the atom. The first publications of Broglie's idea of "matter waves" had drawn little attention from other physicists, but a copy of his doctoral thesis chanced to reach Einstein, whose response was enthusiastic. Einstein stressed the importance of Broglie's work both explicitly and by building further on it.

In 1925, Austrian-born physicist Wolfgang Pauli developed the Pauli exclusion principle, which states that no two electrons around a single nucleus in an atom can occupy the same quantum state simultaneously, as described by four quantum numbers. Pauli made major contributions to quantum mechanics and quantum field theory - he was awarded the 1945 Nobel Prize for Physics for his discovery of the Pauli exclusion principle - as well as solid-state physics, and he successfully hypothesized the existence of the neutrino. In addition to his original work, he wrote masterful syntheses of several areas of physical theory that are considered classics of scientific literature.

$$H(t)\,|\,\psi(t)\rangle = i\hbar\frac{d}{dt}\,|\,\psi(t)\rangle$$

The Schrödinger equation

In 1926 at the age of 39, Austrian theoretical physicist Erwin Schrödinger produced the papers that gave the foundations of quantum wave mechanics. In those papers he described his partial differential equation that is the basic equation of quantum mechanics and bears the same relation to the mechanics of the atom as Newton's equations of motion bear to planetary astronomy. Adopting a proposal made by Louis de Broglie in 1924 that particles of matter have a dual nature and in some situations act like waves, Schrödinger introduced a theory describing the behaviour of such a system by a wave equation that is now known as the Schrödinger equation. The solutions to Schrödinger's equation, unlike the solutions to Newton's equations, are wave functions that can only be related to the probable occurrence of physical events. The readily visualized sequence of events of the planetary orbits of Newton is, in quantum mechanics, replaced by the more abstract notion of probability. (This aspect of the quantum theory made Schrödinger and several other physicists profoundly unhappy, and he devoted much of his later life to formulating philosophical objections to the generally accepted interpretation of the theory that he had done so much to create.)

German theoretical physicist Werner Heisenberg was one of the key creators of quantum mechanics. In 1925, Heisenberg discovered a way to formulate quantum mechanics in terms of matrices. For that discovery, he was awarded the Nobel Prize for Physics for 1932. In 1927 he published his uncertainty principle, upon which he built his philosophy and for which he is best known. Heisenberg was able to demonstrate that if you were studying an electron in an atom you could say where it was (the electron's location) or where it was going (the electron's velocity), but it was impossible to express both at the same time. He also made important contributions to the theories of the hydrodynamics of turbulent flows, the atomic nucleus, ferromagnetism, cosmic rays, and subatomic particles, and he was instrumental in planning the first West German nuclear reactor at Karlsruhe, together with a research reactor in Munich, in 1957. Considerable controversy surrounds his work on atomic research during World War II.

Quantum Chemistry

Some view the birth of quantum chemistry in the discovery of the Schrödinger equation and its application to the hydrogen atom in 1926. However, the 1927 article of Walter Heitler and Fritz London is often recognised as the first milestone in the history of quantum chemistry. This is the first application of quantum mechanics to the diatomic hydrogen molecule, and thus to the phenomenon of the chemical bond. In the following years much progress was accomplished by Edward Teller, Robert S. Mulliken, Max Born, J. Robert Oppenheimer, Linus Pauling, Erich Hückel, Douglas Hartree, Vladimir Aleksandrovich Fock, to cite a few.

Still, skepticism remained as to the general power of quantum mechanics applied to complex chemical systems. The situation around 1930 is described by Paul Dirac:

The underlying physical laws necessary for the mathematical theory of a large part of physics and the whole of chemistry are thus completely known, and the difficulty is only that the exact application of these laws leads to equations much too complicated to be soluble. It therefore becomes desirable that approximate practical methods of applying quantum mechanics should be developed, which can lead to an explanation of the main features of complex atomic systems without too much computation.

Hence the quantum mechanical methods developed in the 1930s and 1940s are often referred to as theoretical molecular or atomic physics to underline the fact that they were more the application of quantum mechanics to chemistry and spectroscopy than answers to chemically relevant questions. In 1951, a milestone article in quantum chemistry is the seminal paper of Clemens C. J. Roothaan on Roothaan equations. It opened the avenue to the solution of the self-consistent field equations for small molecules like hydrogen or nitrogen. Those computations were performed with the help of tables of integrals which were computed on the most advanced computers of the time.

In the 1940s many physicists turned from molecular or atomic physics to nuclear physics (like J. Robert Oppenheimer or Edward Teller). Glenn T. Seaborg was an American nuclear chemist best known for his work on isolating and identifying transuranium elements (those heavier than uranium). He shared the 1951 Nobel Prize for Chemistry with Edwin Mattison McMillan for their independent discoveries of transuranium elements. Seaborgium was named in his honour, making him the only person, along Albert Einstein, for whom a chemical element was named during his lifetime.

Molecular Biology and Biochemistry

By the mid 20th century, in principle, the integration of physics and chemistry was extensive, with chemical properties explained as the result of the electronic structure of the atom; Linus Pauling's book on *The Nature of the Chemical Bond* used the principles of quantum mechanics to deduce bond angles in ever-more complicated molecules. However, though some principles deduced from quantum mechanics were able to predict qualitatively some chemical features for biologically relevant molecules, they were, till the end of the 20th century, more a collection of rules, observations, and recipes than rigorous ab initio quantitative methods.

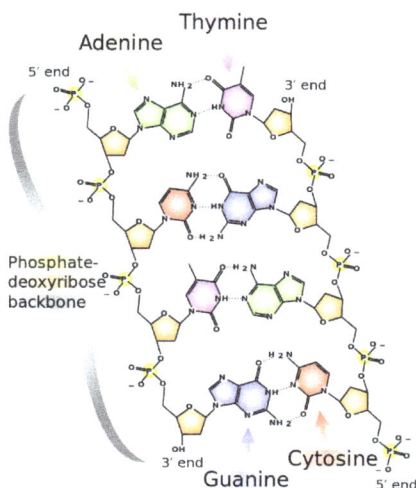

Diagrammatic representation of some key structural features of DNA

This heuristic approach triumphed in 1953 when James Watson and Francis Crick deduced the double helical structure of DNA by constructing models constrained by and informed by the knowledge of the chemistry of the constituent parts and the X-ray diffraction patterns obtained by Rosalind Franklin. This discovery lead to an explosion of research into the biochemistry of life.

In the same year, the Miller–Urey experiment demonstrated that basic constituents of protein, simple amino acids, could themselves be built up from simpler molecules in a simulation of primordial processes on Earth. Though many questions remain about the true nature of the origin of life, this was the first attempt by chemists to study hypothetical processes in the laboratory under controlled conditions.

In 1983 Kary Mullis devised a method for the in-vitro amplification of DNA, known as the polymerase chain reaction (PCR), which revolutionized the chemical processes used in the laboratory to manipulate it. PCR could be used to synthesize specific pieces of DNA and made possible the sequencing of DNA of organisms, which culminated in the huge human genome project.

An important piece in the double helix puzzle was solved by one of Pauling's students Matthew Meselson and Frank Stahl, the result of their collaboration (Meselson–Stahl experiment) has been called as "the most beautiful experiment in biology".

They used a centrifugation technique that sorted molecules according to differences in weight. Because nitrogen atoms are a component of DNA, they were labelled and therefore tracked in replication in bacteria.

Late 20th Century

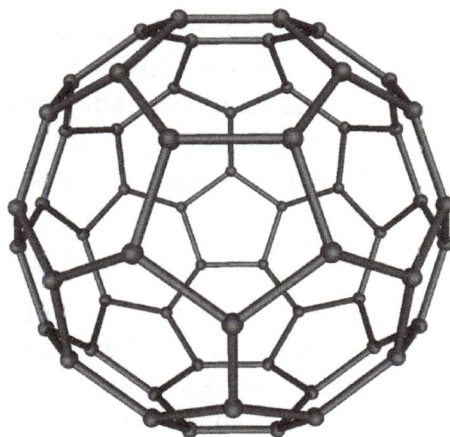

Buckminsterfullerene, C_{60}

In 1970, John Pople developed the Gaussian program greatly easing computational chemistry calculations. In 1971, Yves Chauvin offered an explanation of the reaction mechanism of olefin metathesis reactions. In 1975, Karl Barry Sharpless and his group discovered a stereoselective oxidation reactions including Sharpless epoxidation, Sharpless asymmetric dihydroxylation, and Sharpless oxyamination. In 1985, Harold Kroto, Robert Curl and Richard Smalley discovered fullerenes, a class of large carbon molecules superficially resembling the geodesic dome designed by architect R. Buckminster Fuller. In 1991, Sumio Iijima used electron microscopy to discover a type of cylindrical fullerene known as a carbon nanotube, though earlier work had been done in the field as early as 1951. This material is an important component in the field of nanotechnology. In 1994, Robert A. Holton and his group achieved the first total synthesis of Taxol. In 1995, Eric Cornell and Carl Wieman produced the first Bose–Einstein condensate, a substance that displays quantum mechanical properties on the macroscopic scale.

Mathematics and Chemistry

Classically, before the 20th century, chemistry was defined as the science of the nature of matter and its transformations. It was therefore clearly distinct from physics which was not concerned with such dramatic transformation of matter. Moreover, in contrast to physics, chemistry was not using much of mathematics. Even some were particularly reluctant to use mathematics within chemistry. For example, Auguste Comte wrote in 1830:

Every attempt to employ mathematical methods in the study of chemical questions must be considered profoundly irrational and contrary to the spirit of chemistry.... if mathematical analysis should ever hold a prominent place in chemistry -- an aberration which is happily almost impossible -- it would occasion a rapid and widespread degeneration of that science.

However, in the second part of the 19th century, the situation changed and August Kekulé wrote in 1867:

I rather expect that we shall someday find a mathematico-mechanical explanation for what we now call atoms which will render an account of their properties.

Scope of Chemistry

As understanding of the nature of matter has evolved, so too has the self-understanding of the science of chemistry by its practitioners. This continuing historical process of evaluation includes the categories, terms, aims and scope of chemistry. Additionally, the development of the social institutions and networks which support chemical enquiry are highly significant factors that enable the production, dissemination and application of chemical knowledge.

Chemical Industry

The later part of the nineteenth century saw a huge increase in the exploitation of petroleum extracted from the earth for the production of a host of chemicals and largely replaced the use of whale oil, coal tar and naval stores used previously. Large-scale production and refinement of petroleum provided feedstocks for liquid fuels such as gasoline and diesel, solvents, lubricants, asphalt, waxes, and for the production of many of the common materials of the modern world, such as synthetic fibers, plastics, paints, detergents, pharmaceuticals, adhesives and ammonia as fertilizer and for other uses. Many of these required new catalysts and the utilization of chemical engineering for their cost-effective production.

In the mid-twentieth century, control of the electronic structure of semiconductor materials was made precise by the creation of large ingots of extremely pure single crystals of silicon and germanium. Accurate control of their chemical composition by doping with other elements made the production of the solid state transistor in 1951 and made possible the production of tiny integrated circuits for use in electronic devices, especially computers.

Significant Topics in Chemistry

Matter and atoms are the building blocks of chemistry. The following chapter unfolds its vital aspects in a critical yet systematic manner. It strategically encompasses and incorporates the major components and key concepts of chemistry, providing a complete understanding.

Matter

Before the 20th century, the term matter included ordinary matter composed of atoms and excluded other energy phenomena such as light or sound. This concept of matter may be generalized from atoms to include any objects having mass even when at rest, but this is ill-defined because an object's mass can arise from its (possibly massless) constituents' motion and interaction energies. Thus, matter does not have a universal definition, nor is it a fundamental concept in physics today. Matter is also used loosely as a general term for the substance that makes up all observable physical objects.

All the objects from everyday life that we can bump into, touch or squeeze are composed of atoms. This atomic matter is in turn made up of interacting subatomic particles—usually a nucleus of protons and neutrons, and a cloud of orbiting electrons. Typically, science considers these composite particles matter because they have both rest mass and volume. By contrast, massless particles, such as photons, are not considered matter, because they have neither rest mass nor volume. However, not all particles with rest mass have a classical volume, since fundamental particles such as quarks and leptons (sometimes equated with matter) are considered "point particles" with no effective size or volume. Nevertheless, quarks and leptons together make up "ordinary matter", and their interactions contribute to the effective volume of the composite particles that make up ordinary matter.

Matter exists in *states* (or *phases*): the classical solid, liquid, and gas; as well as the more exotic plasma, Bose–Einstein condensates, fermionic condensates, and quark–gluon plasma.

For much of the history of the natural sciences people have contemplated the exact nature of matter. The idea that matter was built of discrete building blocks, the so-called *particulate theory of matter*, was first put forward by the Greek philosophers Leucippus (~490 BC) and Democritus (~470–380 BC).

Comparison with Mass

Matter should not be confused with mass, as the two are not quite the same in modern physics. For example, mass is a conserved quantity, which means that its value is unchanging through time, within closed systems. However, matter is *not* conserved in such systems, although this is not obvious in ordinary conditions on Earth, where matter is approximately conserved. Still, special relativity shows that matter may disappear by conversion into energy, even inside closed systems, and it can also be created from energy, within such systems. However, because *mass* (like energy) can neither be created nor destroyed, the quantity of mass and the quantity of energy remain the same during a transformation of matter (which represents a certain amount of energy) into non-material (i.e., non-matter) energy. This is also true in the reverse transformation of energy into matter.

Different fields of science use the term matter in different, and sometimes incompatible, ways. Some of these ways are based on loose historical meanings, from a time when there was no reason to distinguish mass and matter. As such, there is no single universally agreed scientific meaning of the word "matter". Scientifically, the term "mass" is well-defined, but "matter" is not. Sometimes in the field of physics "matter" is simply equated with particles that exhibit rest mass (i.e., that cannot travel at the speed of light), such as quarks and leptons. However, in both physics and chemistry, matter exhibits both wave-like and particle-like properties, the so-called wave–particle duality.

Definition

Based on mass, Volume, and Space

The common definition of matter is *anything that has mass and volume (occupies space)*. For example, a car would be said to be made of matter, as it has mass and volume (occupies space).

The observation that matter occupies space goes back to antiquity. However, an explanation for why matter occupies space is recent, and is argued to be a result of the phenomenon described in the Pauli exclusion principle. Two particular examples where the exclusion principle clearly relates matter to the occupation of space are white dwarf stars and neutron stars, discussed further below.

Based on Atoms

A definition of "matter" based on its physical and chemical structure is: *matter is made up of atoms*. As an example, deoxyribonucleic acid molecules (DNA) are matter under this definition because they are made of atoms. This definition can extend to include charged atoms and molecules, so as to include plasmas (gases of ions) and electrolytes (ionic solutions), which are not obviously included in the atoms definition. Alternatively, one can adopt the *protons, neutrons, and electrons* definition.

Based on Protons, neutrons and Electrons

A definition of "matter" more fine-scale than the atoms and molecules definition is: *matter is made up of what atoms and molecules are made of*, meaning anything made of positively charged protons, neutral neutrons, and negatively charged electrons. This definition goes beyond atoms and molecules, however, to include substances made from these building blocks that are *not* simply atoms or molecules, for example white dwarf matter—typically, carbon and oxygen nuclei in a sea of degenerate electrons. At a microscopic level, the constituent "particles" of matter such as protons, neutrons, and electrons obey the laws of quantum mechanics and exhibit wave–particle duality. At an even deeper level, protons and neutrons are made up of quarks and the force fields (gluons) that bind them together.

Based on Quarks and Leptons

Under the "quarks and leptons" definition, the elementary and composite particles made of the quarks (in purple) and leptons (in green) would be matter—while the gauge bosons (in red) would not be matter. However, interaction energy inherent to composite particles (for example, gluons involved in neutrons and protons) contribute to the mass of ordinary matter.

As seen in the above discussion, many early definitions of what can be called *ordinary matter* were based upon its structure or *building blocks*. On the scale of elementary particles, a definition that follows this tradition can be stated as: *ordinary matter is everything that is composed of elementary fermions, namely quarks and leptons.* The connection between these formulations follows.

Leptons (the most famous being the electron), and quarks (of which baryons, such as protons and neutrons, are made) combine to form atoms, which in turn form molecules. Because atoms and molecules are said to be matter, it is natural to phrase the definition as: *ordinary matter is anything that is made of the same things that atoms and molecules are made of.* (However, notice that one also can make from these building blocks matter that is *not* atoms or molecules.) Then, because electrons are leptons, and protons, and neutrons are made of quarks, this definition in turn leads to the definition of matter as being *quarks and leptons*, which are the two types of elementary fermions. Carithers and Grannis state: *Ordinary matter is composed entirely of first-generation particles, namely the [up] and [down] quarks, plus the electron and its neutrino.* (Higher generations particles quickly decay into first-generation particles, and thus are not commonly encountered.)

This definition of ordinary matter is more subtle than it first appears. All the particles that make up ordinary matter (leptons and quarks) are elementary fermions, while all the force carriers are elementary bosons. The W and Z bosons that mediate the weak force are not made of quarks or leptons, and so are not ordinary matter, even if they have mass. In other words, mass is not something that is exclusive to ordinary matter.

The quark–lepton definition of ordinary matter, however, identifies not only the elementary building blocks of matter, but also includes composites made from the constituents (atoms and molecules, for example). Such composites contain an interaction energy that holds the constituents together, and may constitute the bulk of the mass of the composite. As an example, to a great extent, the mass of an atom is simply the sum of the masses of its constituent protons, neutrons and electrons. However, digging deeper, the protons and neutrons are made up of quarks bound together by gluon fields and these gluons fields contribute significantly to the mass of hadrons. In other words, most of what composes the "mass" of ordinary matter is due to the binding energy of quarks within protons and neutrons. For example, the sum of the mass of the three quarks in a nucleon is approximately 12.5 MeV/c^2, which is low compared to the mass of a nucleon (approximately 938 MeV/c^2). The bottom line is that most of the mass of everyday objects comes from the interaction energy of its elementary components.

The Standard Model groups matter particles into three generations, where each generation consists of two quarks and two leptons. The first generation is the *up* and *down* quarks, the *electron* and the *electron neutrino*; the second includes the *charm* and *strange* quarks, the *muon* and the *muon neutrino*; the third generation consists of the *top* and *bottom* quarks and the *tau* and *tau neutrino*. The most natural explanation for this would be that quarks and leptons of higher generations are excited states of the first generations. If this turns out to be the case, it would imply that quarks and leptons are composite particles, rather than elementary particles.

Based on Theories of Relativity

In the context of relativity, mass is not an additive quantity, in the sense that one can add the rest masses of particles in a system to get the total rest mass of the system. Thus, in relativity usually a more general view is that it is not the sum of rest masses, but the energy–momentum tensor that quantifies the amount of matter. This tensor gives the rest mass for the entire system. "Matter" therefore is sometimes considered as anything that contributes to the energy–momentum of a system, that is, anything that is not purely gravity. This view is commonly held in fields that deal with general relativity such as cosmology. In this view, light and other massless particles and fields are part of matter.

The reason for this is that in this definition, electromagnetic radiation (such as light) as well as the energy of electromagnetic fields contributes to the mass of systems, and therefore appears to add matter to them. For example, light radiation (or thermal radiation) trapped inside a box would contribute to the mass of the box, as would any kind of energy inside the box, including the kinetic energy of particles held by the box. Nevertheless, isolated individual particles of light (photons) and the isolated kinetic energy of massive particles, are normally not considered to be *matter*.

A difference between matter and mass therefore may seem to arise when single particles are examined. In such cases, the mass of single photons is zero. For particles with rest mass, such as leptons and quarks, isolation of the particle in a frame where it is not moving, removes its kinetic energy.

A source of definition difficulty in relativity arises from two definitions of mass in common use, one of which is formally equivalent to total energy (and is thus observer dependent), and the other of which is referred to as rest mass or invariant mass and is independent of the observer. Only "rest mass" is loosely equated with matter (since it can be weighed). Invariant mass is usually applied in physics to unbound systems of particles. However, energies which contribute to the "invariant mass" may be weighed also in special circumstances, such as when a system that has invariant mass is confined and has no net momentum (as in the box example above). Thus, a photon with no mass may (confusingly) still add mass to a system in which it is trapped. The same is true of the kinetic energy of particles, which by definition is not part of their rest mass, but which does add rest mass to systems in which these particles reside (an example is the mass added by the motion of gas molecules of a bottle of gas, or by the thermal energy of any hot object).

Since such mass (kinetic energies of particles, the energy of trapped electromagnetic radiation and stored potential energy of repulsive fields) is measured as part of the mass of ordinary *matter* in complex systems, the "matter" status of "massless particles" and fields of force becomes unclear in such systems. These problems contribute to the lack of a rigorous definition of matter in science, although mass is easier to define as the total stress–energy above (this is also what is weighed on a scale, and what is the source of gravity).

Structure

In particle physics, fermions are particles that obey Fermi–Dirac statistics. Fermions can be elementary, like the electron—or composite, like the proton and neutron. In the Standard Model, there are two types of elementary fermions: quarks and leptons, which are discussed next.

Quarks

Quarks are particles of spin-$\frac{1}{2}$, implying that they are fermions. They carry an electric charge of $-\frac{1}{3}$ e (down-type quarks) or $+\frac{2}{3}$ e (up-type quarks). For comparison, an electron has a charge of -1 e. They also carry colour charge, which is the equivalent of the electric charge for the strong interaction. Quarks also undergo radioactive decay, meaning that they are subject to the weak interaction. Quarks are massive particles, and therefore are also subject to gravity.

Quark properties							
name	symbol	spin	electric charge (e)	mass (MeV/c²)	mass comparable to	antiparticle	antiparticle symbol
up-type quarks							
up	u	$\frac{1}{2}$	$+\frac{2}{3}$	1.5 to 3.3	~ 5 electrons	antiup	u
charm	c	$\frac{1}{2}$	$+\frac{2}{3}$	1160 to 1340	~1 proton	anticharm	c
top	t	$\frac{1}{2}$	$+\frac{2}{3}$	169,100 to 173,300	~180 protons or ~1 tungsten atom	antitop	t
down-type quarks							
down	d	$\frac{1}{2}$	$-\frac{1}{3}$	3.5 to 6.0	~10 electrons	antidown	d
strange	s	$\frac{1}{2}$	$-\frac{1}{3}$	70 to 130	~ 200 electrons	antistrange	s
bottom	b	$\frac{1}{2}$	$-\frac{1}{3}$	4130 to 4370	~ 5 protons	antibottom	b

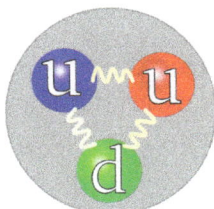

Quark structure of a proton: 2 up quarks and 1 down quark.

Baryonic Matter

Baryons are strongly interacting fermions, and so are subject to Fermi–Dirac statistics. Amongst the baryons are the protons and neutrons, which occur in atomic nuclei, but many other unstable baryons exist as well. The term baryon usually refers to triquarks—particles made of three quarks. "Exotic" baryons made of four quarks and one antiquark are known as the pentaquarks, but their existence is not generally accepted.

Baryonic matter is the part of the universe that is made of baryons (including all atoms). This part of the universe does not include dark energy, dark matter, black holes or various forms of degenerate matter, such as compose white dwarf stars and neutron stars. Microwave light seen by Wilkinson Microwave Anisotropy Probe (WMAP), suggests that only about 4.6% of that part of the universe within range of the best telescopes (that is, matter that may be visible because light could reach us from it), is made of baryonic matter. About 23% is dark matter, and about 72% is dark energy.

As a matter of fact, the great majority of ordinary matter in the universe is unseen, since visible stars and gas inside galaxies and clusters account for less than 10 per cent of the ordinary matter contribution to the mass-energy density of the universe.

A comparison between the white dwarf IK Pegasi B (center), its A-class companion IK Pegasi A (left) and the Sun (right). This white dwarf has a surface temperature of 35,500 K.

Degenerate Matter

In physics, degenerate matter refers to the ground state of a gas of fermions at a temperature near absolute zero. The Pauli exclusion principle requires that only two fermions can occupy a quantum state, one spin-up and the other spin-down. Hence, at zero temperature, the fermions fill up sufficient levels to accommodate all the available fermions—and in the case of many fermions, the maximum kinetic energy (called the *Fermi energy*) and the pressure of the gas becomes very large, and depends on the number of fermions rather than the temperature, unlike normal states of matter.

Degenerate matter is thought to occur during the evolution of heavy stars. The demonstration by Subrahmanyan Chandrasekhar that white dwarf stars have a maximum allowed mass because of the exclusion principle caused a revolution in the theory of star evolution.

Degenerate matter includes the part of the universe that is made up of neutron stars and white dwarfs.

Strange Matter

Strange matter is a particular form of quark matter, usually thought of as a liquid of up, down, and strange quarks. It is contrasted with nuclear matter, which is a liquid of neutrons and protons (which themselves are built out of up and down quarks), and with non-strange quark matter, which is a quark liquid that contains only up and down quarks. At high enough density, strange matter is expected to be color superconducting. Strange matter is hypothesized to occur in the core of neutron stars, or, more speculatively, as isolated droplets that may vary in size from femtometers (strangelets) to kilometers (quark stars).

Two meanings of the Term "Strange Matter"

In particle physics and astrophysics, the term is used in two ways, one broader and the other more specific.

1. The broader meaning is just quark matter that contains three flavors of quarks: up, down, and strange. In this definition, there is a critical pressure and an associated critical density, and when nuclear matter (made of protons and neutrons) is compressed beyond this density, the protons and neutrons dissociate into quarks, yielding quark matter (probably strange matter).

2. The narrower meaning is quark matter that is *more stable than nuclear matter*. The idea that this could happen is the "strange matter hypothesis" of Bodmer and Witten. In this definition, the critical pressure is zero: the true ground state of matter is *always* quark matter. The nuclei that we see in the matter around us, which are droplets of nuclear matter, are actually metastable, and given enough time (or the right external stimulus) would decay into droplets of strange matter, i.e. strangelets.

Leptons

Leptons are particles of spin-$\frac{1}{2}$, meaning that they are fermions. They carry an electric charge of −1 e (charged leptons) or 0 e (neutrinos). Unlike quarks, leptons do not carry colour charge, meaning that they do not experience the strong interaction. Leptons also undergo radioactive decay, meaning that they are subject to the weak interaction. Leptons are massive particles, therefore are subject to gravity.

Lepton properties							
name	symbol	spin	electric charge (e)	mass (MeV/c²)	mass comparable to	antiparticle	antiparticle symbol
charged leptons							
electron	e−	$\frac{1}{2}$	−1	0.5110	1 electron	antielectron	e+
muon	μ−	$\frac{1}{2}$	−1	105.7	~ 200 electrons	antimuon	μ+
tau	τ−	$\frac{1}{2}$	−1	1,777	~ 2 protons	antitau	τ+
neutrinos							
electron neutrino	ν e	$\frac{1}{2}$	0	< 0.000460	< $\frac{1}{1000}$ electron	electron antineutrino	ν e
muon neutrino	ν μ	$\frac{1}{2}$	0	< 0.19	< $\frac{1}{2}$ electron	muon antineutrino	ν μ
tau neutrino	ν τ	$\frac{1}{2}$	0	< 18.2	< 40 electrons	tau antineutrino	ν τ

Phases

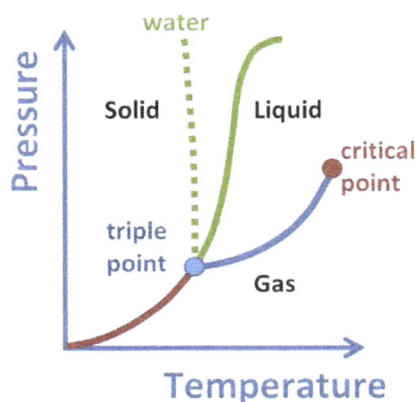

Phase diagram for a typical substance at a fixed volume. Vertical axis is *Pressure*, horizontal axis is *Temperature*. The green line marks the freezing point (above the green line is *solid*, below it is *liquid*) and the blue line the boiling point (above it is *liquid* and below it is *gas*). So, for example, at higher *T*, a higher *P* is necessary to maintain the substance in liquid phase. At the triple point the three phases; liquid, gas and solid; can coexist. Above the critical point there is no detectable difference between the phases. The dotted line shows the anomalous behavior of water: ice melts at constant temperature with increasing pressure.

In bulk, matter can exist in several different forms, or states of aggregation, known as *phases*, depending on ambient pressure, temperature and volume. A phase is a form of matter that has a relatively uniform chemical composition and physical properties (such as density, specific heat, refractive index, and so forth). These phases include the three familiar ones (solids, liquids, and gases), as well as more exotic states of matter (such as plasmas, superfluids, supersolids, Bose–Einstein condensates, ...). A *fluid* may be a liquid, gas or plasma. There are also paramagnetic and ferromagnetic phases of magnetic materials. As conditions change, matter may change from one phase into another. These phenomena are called phase transitions, and are studied in the field of thermodynamics. In nanomaterials, the vastly increased ratio of surface area to volume results in matter that can exhibit properties entirely different from those of bulk material, and not well described by any bulk phase.

Phases are sometimes called *states of matter*, but this term can lead to confusion with thermodynamic states. For example, two gases maintained at different pressures are in different *thermodynamic states* (different pressures), but in the same *phase* (both are gases).

Antimatter

In particle physics and quantum chemistry, antimatter is matter that is composed of the antiparticles of those that constitute ordinary matter. If a particle and its antiparticle come into contact with each other, the two annihilate; that is, they may both be converted into other particles with equal energy in accordance with Einstein's equation $E = mc^2$. These new particles may be high-energy photons (gamma rays) or other particle–antiparticle pairs. The resulting particles are endowed with an amount of kinetic energy equal to the difference between the rest mass of the products of the annihilation and the rest mass of the original particle–antiparticle pair, which is often quite large.

Antimatter is not found naturally on Earth, except very briefly and in vanishingly small quantities (as the result of radioactive decay, lightning or cosmic rays). This is because antimatter that came to exist on Earth outside the confines of a suitable physics laboratory would almost instantly meet the ordinary matter that Earth is made of, and be annihilated. Antiparticles and some stable antimatter (such as antihydrogen) can be made in tiny amounts, but not in enough quantity to do more than test a few of its theoretical properties.

There is considerable speculation both in science and science fiction as to why the observable universe is apparently almost entirely matter, and whether other places are almost entirely antimatter instead. In the early universe, it is thought that matter and antimatter were equally represented, and the disappearance of antimatter requires an asymmetry in physical laws called the charge parity (or CP symmetry) violation. CP symmetry violation can be obtained from the Standard Model, but at this time the apparent asymmetry of matter and antimatter in the visible universe is one of the great unsolved problems in physics. Possible processes by which it came about are explored in more detail under baryogenesis.

Other Types

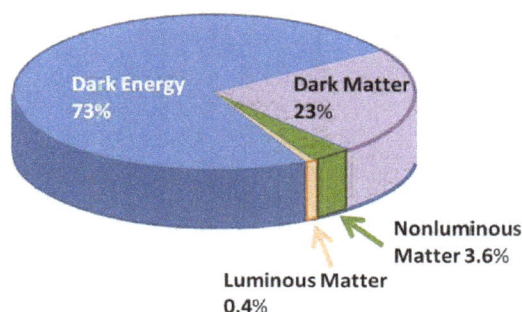

Pie chart showing the fractions of energy in the universe contributed by different sources. *Ordinary matter* is divided into *luminous matter* (the stars and luminous gases and 0.005% radiation) and *nonluminous matter* (intergalactic gas and about 0.1% neutrinos and 0.04% supermassive black holes). Ordinary matter is uncommon. Modeled after Ostriker and Steinhardt. For more information.

Ordinary matter, in the quarks and leptons definition, constitutes about 4% of the energy of the observable universe. The remaining energy is theorized to be due to exotic forms, of which 23% is dark matter and 73% is dark energy.

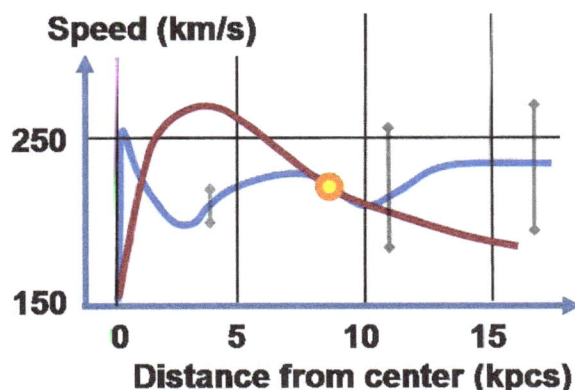

Galaxy rotation curve for the Milky Way. Vertical axis is speed of rotation about the galactic center. Horizontal axis is distance from the galactic center. The sun is marked with a yellow ball. The observed curve of speed of rotation is blue. The predicted curve based upon stellar mass and gas in the Milky Way is red. The difference is due to dark matter or perhaps a modification of the law of gravity. Scatter in observations is indicated roughly by gray bars.

Dark Matter

In astrophysics and cosmology, dark matter is matter of unknown composition that does not emit or reflect enough electromagnetic radiation to be observed directly, but whose presence can be inferred from gravitational effects on visible matter. Observational evidence of the early universe and the big bang theory require that this matter have energy and mass, but is not composed of either elementary fermions (as above) OR gauge bosons. The commonly accepted view is that most of the dark matter is non-baryonic in nature. As such, it is composed of particles as yet unobserved in the laboratory. Perhaps they are supersymmetric particles, which are not Standard Model particles, but relics formed at very high energies in the early phase of the universe and still floating about.

Dark Energy

In cosmology, dark energy is the name given to the antigravitating influence that is accelerating the rate of expansion of the universe. It is known not to be composed of known particles like protons, neutrons or electrons, nor of the particles of dark matter, because these all gravitate.

Fully 70% of the matter density in the universe appears to be in the form of dark energy. Twenty-six percent is dark matter. Only 4% is ordinary matter. So less than 1 part in 20 is made out of matter we have observed experimentally or described in the standard model of particle physics. Of the other 96%, apart from the properties just mentioned, we know absolutely nothing.

— Lee Smolin: The Trouble with Physics

Exotic Matter

Exotic matter is a hypothetical concept of particle physics. It covers any material that violates one or more classical conditions or is not made of known baryonic particles. Such materials would possess qualities like negative mass or being repelled rather than attracted by gravity.

Historical Development

Antiquity (c. 610 BC–c. 322 BC)

The pre-Socratics were among the first recorded speculators about the underlying nature of the visible world. Thales (c. 624 BC–c. 546 BC) regarded water as the fundamental material of the world. Anaximander (c. 610 BC–c. 546 BC) posited that the basic material was wholly characterless or limitless: the Infinite (*apeiron*). Anaximenes (flourished 585 BC, d. 528 BC) posited that the basic stuff was *pneuma* or air. Heraclitus (c. 535–c. 475 BC) seems to say the basic element is fire, though perhaps he means that all is change. Empedocles (c. 490–430 BC) spoke of four elements of which everything was made: earth, water, air, and fire. Meanwhile, Parmenides argued that change does not exist, and Democritus argued that everything is composed of minuscule, inert bodies of all shapes called atoms, a philosophy called atomism. All of these notions had deep philosophical problems.

Aristotle (384 BC – 322 BC) was the first to put the conception on a sound philosophical basis, which he did in his natural philosophy, especially in *Physics* book I. He adopted as reasonable suppositions the four Empedoclean elements, but added a fifth, aether. Nevertheless, these elements are not basic in Aristotle's mind. Rather they, like everything else in the visible world, are composed of the basic *principles* matter and form.

For my definition of matter is just this—the primary substratum of each thing, from which it comes to be without qualification, and which persists in the result.

—Aristotle, Physics I:9:192a32

The word Aristotle uses for matter, ὕλη (*hyle* or *hule*), can be literally translated as wood or timber, that is, "raw material" for building. Indeed, Aristotle's conception of matter is intrinsically linked to something being made or composed. In other words, in contrast to the early modern conception of matter as simply occupying space, matter for Aristotle is definitionally linked to process or change: matter is what underlies a change of substance. For example, a horse eats grass: the horse changes the grass into itself; the grass as such does not persist in the horse, but some aspect of it—its matter— does. The matter is not specifically described (e.g., as atoms), but consists of whatever persists in the change of substance from grass to horse. Matter in this understanding does not exist independently (i.e., as a substance), but exists interdependently (i.e., as a "principle") with form and only insofar as it underlies change. It can be helpful to conceive of the relationship of matter and form as very similar to that between parts and whole. For Aristotle, matter as such can only *receive* actuality from form; it has no activity or actuality in itself, similar to the way that parts as such only have their existence *in* a whole (otherwise they would be independent wholes).

Seventeenth and Eighteenth Centuries

René Descartes (1596–1650) originated the modern conception of matter. He was primarily a geometer. Instead of, like Aristotle, deducing the existence of matter from the physical reality of change, Descartes arbitrarily postulated matter to be an abstract, mathematical substance that occupies space:

So, extension in length, breadth, and depth, constitutes the nature of bodily substance; and thought constitutes the nature of thinking substance. And everything else attributable to body presupposes extension, and is only a mode of extended

—René Descartes, Principles of Philosophy

For Descartes, matter has only the property of extension, so its only activity aside from locomotion is to exclude other bodies: this is the mechanical philosophy. Descartes makes an absolute distinction between mind, which he defines as unextended, thinking substance, and matter, which he defines as unthinking, extended substance. They are independent things. In contrast, Aristotle defines matter and the formal/forming principle as complementary *principles* that together compose one independent thing (substance). In short, Aristotle defines matter (roughly speaking) as what things are actually made of (with a *potential* independent existence), but Descartes elevates matter to an actual independent thing in itself.

The continuity and difference between Descartes' and Aristotle's conceptions is noteworthy. In both conceptions, matter is passive or inert. In the respective conceptions matter has different relationships to intelligence. For Aristotle, matter and intelligence (form) exist together in an interdependent relationship, whereas for Descartes, matter and intelligence (mind) are definitionally opposed, independent substances.

Descartes' justification for restricting the inherent qualities of matter to extension is its permanence, but his real criterion is not permanence (which equally applied to color and resistance), but his desire to use geometry to explain all material properties. Like Descartes, Hobbes, Boyle, and Locke argued that the inherent properties of bodies were limited to extension, and that so-called secondary qualities, like color, were only products of human perception.

Isaac Newton (1643–1727) inherited Descartes' mechanical conception of matter. In the third of his "Rules of Reasoning in Philosophy", Newton lists the universal qualities of matter as "extension, hardness, impenetrability, mobility, and inertia". Similarly in *Optics* he conjectures that God created matter as "solid, massy, hard, impenetrable, movable particles", which were "...even so very hard as never to wear or break in pieces". The "primary" properties of matter were amenable to mathematical description, unlike "secondary" qualities such as color or taste. Like Descartes, Newton rejected the essential nature of secondary qualities.

Newton developed Descartes' notion of matter by restoring to matter intrinsic properties in addition to extension (at least on a limited basis), such as mass. Newton's use of gravitational force, which worked "at a distance", effectively repudiated Descartes' mechanics, in which interactions happened exclusively by contact.

Though Newton's gravity would seem to be a *power* of bodies, Newton himself did not admit it to be an *essential* property of matter. Carrying the logic forward more consistently, Joseph Priestley (1733-1804) argued that corporeal properties transcend contact mechanics: chemical properties require the *capacity* for attraction. He argued matter has other inherent powers besides the so-called primary qualities of Descartes, et al.

Nineteenth and Twentieth Centuries

Since Priestley's time, there has been a massive expansion in knowledge of the constituents of the material world (viz., molecules, atoms, subatomic particles), but there has been no further development in the *definition* of matter. Rather the question has been set aside. Noam Chomsky (born 1928) summarizes the situation that has prevailed since that time:

What is the concept of body that finally emerged?[...] The answer is that there is no clear and definite conception of body.[...] Rather, the material world is whatever we discover it to be, with whatever properties it must be assumed to have for the purposes of explanatory theory. Any intelligible theory that offers genuine explanations and that can be assimilated to the core notions of physics becomes part of the theory of the material world, part of our account of body. If we have such a theory in some domain, we seek to assimilate it to the core notions of physics, perhaps modifying these notions as we carry out this enterprise.

—*Noam Chomsky, Language and problems of knowledge: the Managua lectures, p. 144*

So matter is whatever physics studies and the object of study of physics is matter: there is no independent general definition of matter, apart from its fitting into the methodology of measurement and controlled experimentation. In sum, the boundaries between what constitutes matter and everything else remains as vague as the demarcation problem of delimiting science from everything else.

In the 19th century, following the development of the periodic table, and of atomic theory, atoms were seen as being the fundamental constituents of matter; atoms formed molecules and compounds.

The common definition in terms of occupying space and having mass is in contrast with most physical and chemical definitions of matter, which rely instead upon its structure and upon attributes not necessarily related to volume and mass. At the turn of the nineteenth century, the knowledge of matter began a rapid evolution.

Aspects of the Newtonian view still held sway. James Clerk Maxwell discussed matter in his work *Matter and Motion*. He carefully separates "matter" from space and time, and defines it in terms of the object referred to in Newton's first law of motion.

However, the Newtonian picture was not the whole story. In the 19th century, the term "matter" was actively discussed by a host of scientists and philosophers, and a brief outline can be found in Levere. A textbook discussion from 1870 suggests matter is what is made up of atoms:

Three divisions of matter are recognized in science: masses, molecules and atoms. A Mass of matter is any portion of matter appreciable by the senses. A Molecule is the smallest particle of matter into which a body can be divided without losing its identity. An Atom is a still smaller particle produced by division of a molecule.

Rather than simply having the attributes of mass and occupying space, matter was held to have chemical and electrical properties. In 1909 the famous physicist J. J. Thomson (1856-1940) wrote about the "constitution of matter" and was concerned with the possible connection between matter and electrical charge.

There is an entire literature concerning the "structure of matter", ranging from the "electrical structure" in the early 20th century, to the more recent "quark structure of matter", introduced today with the remark: *Understanding the quark structure of matter has been one of the most important advances in contemporary physics*. In this connection, physicists speak of *matter fields*, and speak of particles as "quantum excitations of a mode of the matter field". And here is a quote from de Sabbata and Gasperini: "With the word "matter" we denote, in this context, the sources of the interactions, that is spinor fields (like quarks and leptons), which are believed to be the fundamental components of matter, or scalar fields, like the Higgs particles, which are used to introduced mass in a gauge theory (and that, however, could be composed of more fundamental fermion fields)."

In the late 19th century with the discovery of the electron, and in the early 20th century, with the discovery of the atomic nucleus, and the birth of particle physics, matter was seen as made up of electrons, protons and neutrons interacting to form atoms. Today, we know that even protons and neutrons are not indivisible, they can be divided into quarks, while electrons are part of a particle family called leptons. Both quarks and leptons are elementary particles, and are currently seen as being the fundamental constituents of matter.

These quarks and leptons interact through four fundamental forces: gravity, electromagnetism, weak interactions, and strong interactions. The Standard Model of particle physics is currently the best explanation for all of physics, but despite decades of efforts, gravity cannot yet be accounted for at the quantum level; it is only described by classical physics. Interactions between quarks and leptons are the result of an exchange of force-carrying particles (such as photons) between quarks and leptons. The force-carrying particles are not themselves building blocks. As one consequence, mass and energy (which cannot be created or destroyed) cannot always be related to matter (which can be created out of non-matter particles such as photons, or even out of pure energy, such as kinetic energy). Force carriers are usually not considered matter: the carriers of the electric force (photons) possess energy and the carriers of the weak force (W and Z bosons) are massive, but neither are considered matter either. However, while these particles are not considered matter, they do contribute to the total mass of atoms, subatomic particles, and all systems that contain them.

Summary

The modern conception of matter has been refined many times in history, in light of the improvement in knowledge of just *what* the basic building blocks are, and in how they interact. The term "matter" is used throughout physics in a bewildering variety of contexts: for example, one refers to "condensed matter physics", "elementary matter", "partonic" matter, "dark" matter, "anti"-matter, "strange" matter, and "nuclear" matter. In discussions of matter and antimatter, normal matter has been referred to by Alfvén as *koinomatter* (Gk. *common matter*). It is fair to say that in physics, there is no broad consensus as to a general definition of matter, and the term "matter" usually is used in conjunction with a specifying modifier.

The history of the concept of matter is a history of the fundamental *length scales* used to define matter. Different building blocks apply depending upon whether one defines matter on an atomic or elementary particle level. One may use a definition that matter is atoms, or that matter is hadrons, or that matter is leptons and quarks depending upon the scale at which one wishes to define matter.

These quarks and leptons interact through four fundamental forces: gravity, electromagnetism, weak interactions, and strong interactions. The Standard Model of particle physics is currently the best explanation for all of physics, but despite decades of efforts, gravity cannot yet be accounted for at the quantum level; it is only described by classical physics.

Atom

An atom is the smallest constituent unit of ordinary matter that has the properties of a chemical element. Every solid, liquid, gas, and plasma is composed of neutral or ionized atoms. Atoms are very small; typical sizes are around 100 pm (a ten-billionth of a meter, in the short scale). However, atoms do not have well-defined boundaries, and there are different ways to define their size that give different but close values.

Atoms are small enough that attempting to predict their behavior using classical physics - as if they were billiard balls, for example - gives noticeably incorrect predictions due to quantum effects.

Through the development of physics, atomic models have incorporated quantum principles to better explain and predict the behavior.

Every atom is composed of a nucleus and one or more electrons bound to the nucleus. The nucleus is made of one or more protons and typically a similar number of neutrons. Protons and neutrons are called nucleons. More than 99.94% of an atom's mass is in the nucleus. The protons have a positive electric charge, the electrons have a negative electric charge, and the neutrons have no electric charge. If the number of protons and electrons are equal, that atom is electrically neutral. If an atom has more or fewer electrons than protons, then it has an overall negative or positive charge, respectively, and it is called an ion.

The electrons of an atom are attracted to the protons in an atomic nucleus by this electromagnetic force. The protons and neutrons in the nucleus are attracted to each other by a different force, the nuclear force, which is usually stronger than the electromagnetic force repelling the positively charged protons from one another. Under certain circumstances the repelling electromagnetic force becomes stronger than the nuclear force, and nucleons can be ejected from the nucleus, leaving behind a different element: nuclear decay resulting in nuclear transmutation.

The number of protons in the nucleus defines to what chemical element the atom belongs: for example, all copper atoms contain 29 protons. The number of neutrons defines the isotope of the element. The number of electrons influences the magnetic properties of an atom. Atoms can attach to one or more other atoms by chemical bonds to form chemical compounds such as molecules. The ability of atoms to associate and dissociate is responsible for most of the physical changes observed in nature, and is the subject of the discipline of chemistry.

History of Atomic Theory

Atoms in Philosophy

The idea that matter is made up of discrete units is a very old idea, appearing in many ancient cultures such as Greece and India. The word "atom" was coined by ancient Greek philosophers. However, these ideas were founded in philosophical and theological reasoning rather than evidence and experimentation. As a result, their views on what atoms look like and how they behave were incorrect. They also could not convince everybody, so atomism was but one of a number of competing theories on the nature of matter. It was not until the 19th century that the idea was embraced and refined by scientists, when the blossoming science of chemistry produced discoveries that only the concept of atoms could explain.

First Evidence-based Theory

In the early 1800s, John Dalton used the concept of atoms to explain why elements always react in ratios of small whole numbers (the law of multiple proportions). For instance, there are two types of tin oxide: one is 88.1% tin and 11.9% oxygen and the other is 78.7% tin and 21.3% oxygen (tin(II) oxide and tin dioxide respectively). This means that 100g of tin will combine either with 13.5g or 27g of oxygen. 13.5 and 27 form a ratio of 1:2, a ratio of small whole numbers. This common pattern in chemistry suggested to Dalton that elements react in whole number multiples of discrete units—in other words, atoms. In the case of tin oxides, one tin atom will combine with either one or two oxygen atoms.

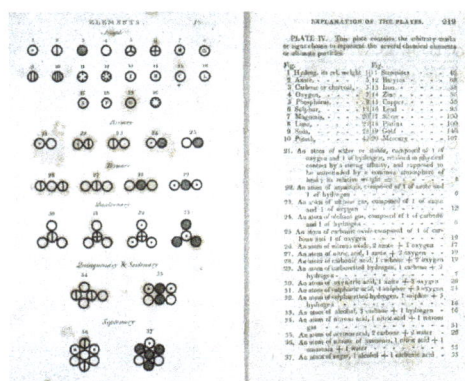

Various atoms and molecules as depicted in John Dalton's *A New System of Chemical Philosophy* (1808).

Dalton also believed atomic theory could explain why water absorbs different gases in different proportions. For example, he found that water absorbs carbon dioxide far better than it absorbs nitrogen. Dalton hypothesized this was due to the differences between the masses and configurations of the gases' respective particles, and carbon dioxide molecules (CO_2) are heavier and larger than nitrogen molecules (N_2).

Brownian Motion

In 1827, botanist Robert Brown used a microscope to look at dust grains floating in water and discovered that they moved about erratically, a phenomenon that became known as "Brownian motion". This was thought to be caused by water molecules knocking the grains about. In 1905 Albert Einstein proved the reality of these molecules and their motions by producing the first Statistical physics analysis of Brownian motion. French physicist Jean Perrin used Einstein's work to experimentally determine the mass and dimensions of atoms, thereby conclusively verifying Dalton's atomic theory.

Discovery of the Electron

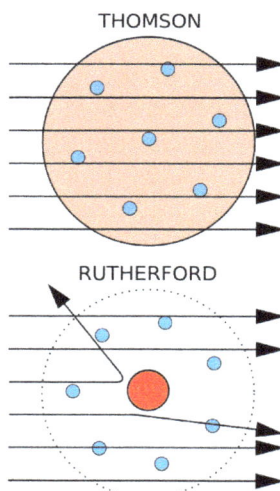

The **Geiger–Marsden experiment**
Top: Expected results: alpha particles passing through the plum pudding model of the atom with negligible deflection.
Bottom: Observed results: a small portion of the particles were deflected by the concentrated positive charge of the nucleus.

The physicist J. J. Thomson measured the mass of cathode rays, showing they were made of particles, but were around 1800 times lighter than the lightest atom, hydrogen. Therefore, they were not atoms, but a new particle, the first *subatomic* particle to be discovered, which he originally called "*corpuscle*" but was later named *electron*, after particles postulated by George Johnstone Stoney in 1874. He also showed they were identical to particles given off by photoelectric and radioactive materials. It was quickly recognized that they are the particles that carry electric currents in metal wires, and carry the negative electric charge within atoms. Thomson was given the 1906 Nobel Prize in Physics for this work. Thus he overturned the belief that atoms are the indivisible, ultimate particles of matter. Thomson also incorrectly postulated that the low mass, negatively charged electrons were distributed throughout the atom in a uniform sea of positive charge. This became known as the plum pudding model.

Discovery of the Nucleus

In 1909, Hans Geiger and Ernest Marsden, under the direction of Ernest Rutherford, bombarded a metal foil with alpha particles to observe how they scattered. They expected all the alpha particles to pass straight through with little deflection, because Thomson's model said that the charges in the atom are so diffuse that their electric fields could not affect the alpha particles much. However, Geiger and Marsden spotted alpha particles being deflected by angles greater than 90°, which was supposed to be impossible according to Thomson's model. To explain this, Rutherford proposed that the positive charge of the atom is concentrated in a tiny nucleus at the center of the atom.

Discovery of Isotopes

While experimenting with the products of radioactive decay, in 1913 radiochemist Frederick Soddy discovered that there appeared to be more than one type of atom at each position on the periodic table. The term isotope was coined by Margaret Todd as a suitable name for different atoms that belong to the same element. J.J. Thomson created a technique for separating atom types through his work on ionized gases, which subsequently led to the discovery of stable isotopes.

Bohr Model

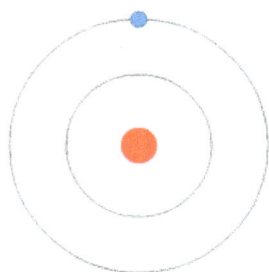

The Bohr model of the atom, with an electron making instantaneous "quantum leaps" from one orbit to another. This model is obsolete.

In 1913 the physicist Niels Bohr proposed a model in which the electrons of an atom were assumed to orbit the nucleus but could only do so in a finite set of orbits, and could jump between these orbits only in discrete changes of energy corresponding to absorption or radiation of a photon. This quantization was used to explain why the electrons orbits are stable (given that normally, charges in acceleration, including circular motion, lose kinetic energy which is emitted as electromagnetic radiation) and why elements absorb and emit electromagnetic radiation in discrete spectra.

Later in the same year Henry Moseley provided additional experimental evidence in favor of Niels Bohr's theory. These results refined Ernest Rutherford's and Antonius Van den Broek's model, which proposed that the atom contains in its nucleus a number of positive nuclear charges that is equal to its (atomic) number in the periodic table. Until these experiments, atomic number was not known to be a physical and experimental quantity. That it is equal to the atomic nuclear charge remains the accepted atomic model today.

Chemical Bonding Explained

Chemical bonds between atoms were now explained, by Gilbert Newton Lewis in 1916, as the interactions between their constituent electrons. As the chemical properties of the elements were known to largely repeat themselves according to the periodic law, in 1919 the American chemist Irving Langmuir suggested that this could be explained if the electrons in an atom were connected or clustered in some manner. Groups of electrons were thought to occupy a set of electron shells about the nucleus.

Further Developments in Quantum Physics

The Stern–Gerlach experiment of 1922 provided further evidence of the quantum nature of the atom. When a beam of silver atoms was passed through a specially shaped magnetic field, the beam was split based on the direction of an atom's angular momentum, or spin. As this direction is random, the beam could be expected to spread into a line. Instead, the beam was split into two parts, depending on whether the atomic spin was oriented up or down.

In 1924, Louis de Broglie proposed that all particles behave to an extent like waves. In 1926, Erwin Schrödinger used this idea to develop a mathematical model of the atom that described the electrons as three-dimensional waveforms rather than point particles. A consequence of using waveforms to describe particles is that it is mathematically impossible to obtain precise values for both the position and momentum of a particle at a given point in time; this became known as the uncertainty principle, formulated by Werner Heisenberg in 1926. In this concept, for a given accuracy in measuring a position one could only obtain a range of probable values for momentum, and vice versa. This model was able to explain observations of atomic behavior that previous models could not, such as certain structural and spectral patterns of atoms larger than hydrogen. Thus, the planetary model of the atom was discarded in favor of one that described atomic orbital zones around the nucleus where a given electron is most likely to be observed.

Discovery of the Neutron

The development of the mass spectrometer allowed the mass of atoms to be measured with increased accuracy. The device uses a magnet to bend the trajectory of a beam of ions, and the amount of deflection is determined by the ratio of an atom's mass to its charge. The chemist Francis William Aston used this instrument to show that isotopes had different masses. The atomic mass of these isotopes varied by integer amounts, called the whole number rule. The explanation for these different isotopes awaited the discovery of the neutron, an uncharged particle with a mass similar to the proton, by the physicist James Chadwick in 1932. Isotopes were then explained as elements with the same number of protons, but different numbers of neutrons within the nucleus.

Fission, High-Energy Physics and Condensed Matter

In 1938, the German chemist Otto Hahn, a student of Rutherford, directed neutrons onto uranium atoms expecting to get transuranium elements. Instead, his chemical experiments showed barium as a product. A year later, Lise Meitner and her nephew Otto Frisch verified that Hahn's result were the first experimental *nuclear fission*. In 1944, Hahn received the Nobel prize in chemistry. Despite Hahn's efforts, the contributions of Meitner and Frisch were not recognized.

In the 1950s, the development of improved particle accelerators and particle detectors allowed scientists to study the impacts of atoms moving at high energies. Neutrons and protons were found to be hadrons, or composites of smaller particles called quarks. The standard model of particle physics was developed that so far has successfully explained the properties of the nucleus in terms of these sub-atomic particles and the forces that govern their interactions.

Structure

Subatomic Particles

Though the word *atom* originally denoted a particle that cannot be cut into smaller particles, in modern scientific usage the atom is composed of various subatomic particles. The constituent particles of an atom are the electron, the proton and the neutron; all three are fermions. However, the hydrogen-1 atom has no neutrons and the hydron ion has no electrons.

The electron is by far the least massive of these particles at 9.11×10^{-31} kg, with a negative electrical charge and a size that is too small to be measured using available techniques. It is the lightest particle with a positive rest mass measured. Under ordinary conditions, electrons are bound to the positively charged nucleus by the attraction created from opposite electric charges. If an atom has more or fewer electrons than its atomic number, then it becomes respectively negatively or positively charged as a whole; a charged atom is called an ion. Electrons have been known since the late 19th century, mostly thanks to J.J. Thomson.

Protons have a positive charge and a mass 1,836 times that of the electron, at 1.6726×10^{-27} kg. The number of protons in an atom is called its atomic number. Ernest Rutherford (1919) observed that nitrogen under alpha-particle bombardment ejects what appeared to be hydrogen nuclei. By 1920 he had accepted that the hydrogen nucleus is a distinct particle within the atom and named it proton.

Neutrons have no electrical charge and have a free mass of 1,839 times the mass of the electron, or 1.6929×10^{-27} kg, the heaviest of the three constituent particles, but it can be reduced by the nuclear binding energy. Neutrons and protons (collectively known as nucleons) have comparable dimensions—on the order of 2.5×10^{-15} m—although the ‹surface› of these particles is not sharply defined. The neutron was discovered in 1932 by the English physicist James Chadwick.

In the Standard Model of physics, electrons are truly elementary particles with no internal structure. However, both protons and neutrons are composite particles composed of elementary particles called quarks. There are two types of quarks in atoms, each having a fractional electric charge. Protons are composed of two up quarks (each with charge +2/3) and one down quark (with a charge of −1/3). Neutrons consist of one up quark and two down quarks. This distinction accounts for the difference in mass and charge between the two particles.

The quarks are held together by the strong interaction (or strong force), which is mediated by gluons. The protons and neutrons, in turn, are held to each other in the nucleus by the nuclear force, which is a residuum of the strong force that has somewhat different range-properties. The gluon is a member of the family of gauge bosons, which are elementary particles that mediate physical forces.

Nucleus

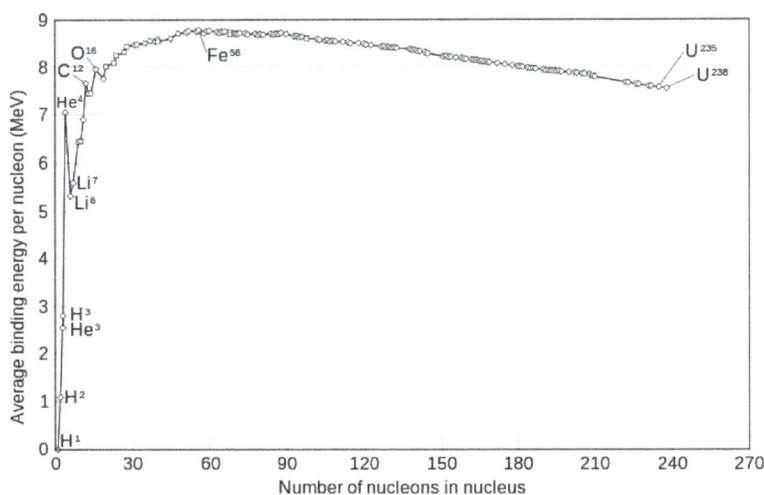

The binding energy needed for a nucleon to escape the nucleus, for various isotopes

All the bound protons and neutrons in an atom make up a tiny atomic nucleus, and are collectively called nucleons. The radius of a nucleus is approximately equal to $1.07 \sqrt[3]{A}$ fm, where A is the total number of nucleons. This is much smaller than the radius of the atom, which is on the order of 10^5 fm. The nucleons are bound together by a short-ranged attractive potential called the residual strong force. At distances smaller than 2.5 fm this force is much more powerful than the electrostatic force that causes positively charged protons to repel each other.

Atoms of the same element have the same number of protons, called the atomic number. Within a single element, the number of neutrons may vary, determining the isotope of that element. The total number of protons and neutrons determine the nuclide. The number of neutrons relative to the protons determines the stability of the nucleus, with certain isotopes undergoing radioactive decay.

The proton, the electron, and the neutron are classified as fermions. Fermions obey the Pauli exclusion principle which prohibits *identical* fermions, such as multiple protons, from occupying the same quantum state at the same time. Thus, every proton in the nucleus must occupy a quantum state different from all other protons, and the same applies to all neutrons of the nucleus and to all electrons of the electron cloud. However, a proton and a neutron are allowed to occupy the same quantum state.

For atoms with low atomic numbers, a nucleus that has more neutrons than protons tends to drop to a lower energy state through radioactive decay so that the neutron–proton ratio is closer to one. However, as the atomic number increases, a higher proportion of neutrons is required to offset the mutual repulsion of the protons. Thus, there are no stable nuclei with equal proton and neutron numbers above atomic number $Z = 20$ (calcium) and as Z increases, the neutron–proton ratio of stable isotopes increases. The stable isotope with the highest proton–neutron ratio is lead-208 (about 1.5).

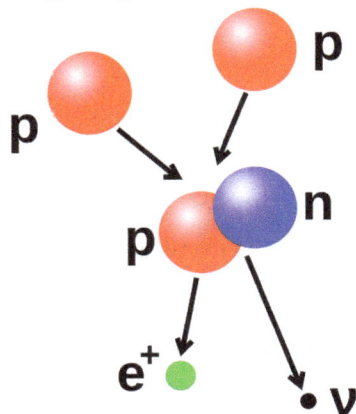

Illustration of a nuclear fusion process that forms a deuterium nucleus, consisting of a proton and a neutron, from two protons. A positron (e^+)—an antimatter electron—is emitted along with an electron neutrino.

The number of protons and neutrons in the atomic nucleus can be modified, although this can require very high energies because of the strong force. Nuclear fusion occurs when multiple atomic particles join to form a heavier nucleus, such as through the energetic collision of two nuclei. For example, at the core of the Sun protons require energies of 3–10 keV to overcome their mutual repulsion—the coulomb barrier—and fuse together into a single nucleus. Nuclear fission is the opposite process, causing a nucleus to split into two smaller nuclei—usually through radioactive decay. The nucleus can also be modified through bombardment by high energy subatomic particles or photons. If this modifies the number of protons in a nucleus, the atom changes to a different chemical element.

If the mass of the nucleus following a fusion reaction is less than the sum of the masses of the separate particles, then the difference between these two values can be emitted as a type of usable energy (such as a gamma ray, or the kinetic energy of a beta particle), as described by Albert Einstein's mass–energy equivalence formula, $E = mc^2$, where m is the mass loss and c is the speed of light. This deficit is part of the binding energy of the new nucleus, and it is the non-recoverable loss of the energy that causes the fused particles to remain together in a state that requires this energy to separate.

The fusion of two nuclei that create larger nuclei with lower atomic numbers than iron and nickel—a total nucleon number of about 60—is usually an exothermic process that releases more energy than is required to bring them together. It is this energy-releasing process that makes nuclear fusion in stars a self-sustaining reaction. For heavier nuclei, the binding energy per nucleon in the nucleus begins to decrease. That means fusion processes producing nuclei that have atomic numbers higher than about 26, and atomic masses higher than about 60, is an endothermic process. These more massive nuclei can not undergo an energy-producing fusion reaction that can sustain the hydrostatic equilibrium of a star.

Electron Cloud

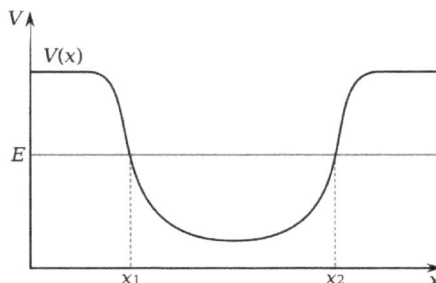

A potential well, showing, according to classical mechanics, the minimum energy $V(x)$ needed to reach each position x. Classically, a particle with energy E is constrained to a range of positions between x_1 and x_2.

The electrons in an atom are attracted to the protons in the nucleus by the electromagnetic force. This force binds the electrons inside an electrostatic potential well surrounding the smaller nucleus, which means that an external source of energy is needed for the electron to escape. The closer an electron is to the nucleus, the greater the attractive force. Hence electrons bound near the center of the potential well require more energy to escape than those at greater separations.

Electrons, like other particles, have properties of both a particle and a wave. The electron cloud is a region inside the potential well where each electron forms a type of three-dimensional standing wave—a wave form that does not move relative to the nucleus. This behavior is defined by an atomic orbital, a mathematical function that characterises the probability that an electron appears to be at a particular location when its position is measured. Only a discrete (or quantized) set of these orbitals exist around the nucleus, as other possible wave patterns rapidly decay into a more stable form. Orbitals can have one or more ring or node structures, and differ from each other in size, shape and orientation.

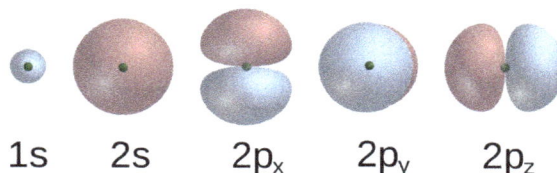

1s   2s   2p$_x$ 2p$_y$ 2p$_z$

Wave functions of the first five atomic orbitals. The three 2p orbitals each display a single angular node that has an orientation and a minimum at the center.

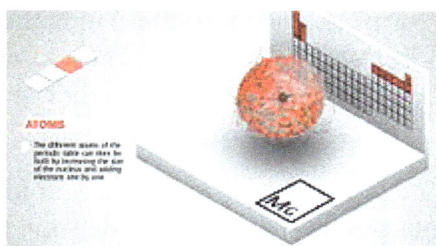

How atoms are constructed from electron orbitals and link to the periodic table

Each atomic orbital corresponds to a particular energy level of the electron. The electron can change its state to a higher energy level by absorbing a photon with sufficient energy to boost it into the new quantum state. Likewise, through spontaneous emission, an electron in a higher energy state can drop to a lower energy state while radiating the excess energy as a photon. These characteristic energy values, defined by the differences in the energies of the quantum states, are responsible for atomic spectral lines.

The amount of energy needed to remove or add an electron—the electron binding energy—is far less than the binding energy of nucleons. For example, it requires only 13.6 eV to strip a ground-state electron from a hydrogen atom, compared to 2.23 *million* eV for splitting a deuterium nucleus. Atoms are electrically neutral if they have an equal number of protons and electrons. Atoms that have either a deficit or a surplus of electrons are called ions. Electrons that are farthest from the nucleus may be transferred to other nearby atoms or shared between atoms. By this mechanism, atoms are able to bond into molecules and other types of chemical compounds like ionic and covalent network crystals.

Properties

Nuclear Properties

By definition, any two atoms with an identical number of *protons* in their nuclei belong to the same chemical element. Atoms with equal numbers of protons but a different number of *neutrons* are different isotopes of the same element. For example, all hydrogen atoms admit exactly one proton, but isotopes exist with no neutrons (hydrogen-1, by far the most common form, also called protium), one neutron (deuterium), two neutrons (tritium) and more than two neutrons. The known elements form a set of atomic numbers, from the single proton element hydrogen up to the 118-proton element ununoctium. All known isotopes of elements with atomic numbers greater than 82 are radioactive.

About 339 nuclides occur naturally on Earth, of which 254 (about 75%) have not been observed to decay, and are referred to as "stable isotopes". However, only 90 of these nuclides are stable to all decay, even in theory. Another 164 (bringing the total to 254) have not been observed to decay, even though in theory it is energetically possible. These are also formally classified as "stable". An additional 34 radioactive nuclides have half-lives longer than 80 million years, and are long-lived enough to be present from the birth of the solar system. This collection of 288 nuclides are known as primordial nuclides. Finally, an additional 51 short-lived nuclides are known to occur naturally, as daughter products of primordial nuclide decay (such as radium from uranium), or else as products of natural energetic processes on Earth, such as cosmic ray bombardment (for example, carbon-14).

For 80 of the chemical elements, at least one stable isotope exists. As a rule, there is only a handful of stable isotopes for each of these elements, the average being 3.2 stable isotopes per element. Twenty-six elements have only a single stable isotope, while the largest number of stable isotopes observed for any element is ten, for the element tin. Elements 43, 61, and all elements numbered 83 or higher have no stable isotopes.

Stability of isotopes is affected by the ratio of protons to neutrons, and also by the presence of certain "magic numbers" of neutrons or protons that represent closed and filled quantum shells. These quantum shells correspond to a set of energy levels within the shell model of the nucleus; filled shells, such as the filled shell of 50 protons for tin, confers unusual stability on the nuclide. Of the 254 known stable nuclides, only four have both an odd number of protons *and* odd number of neutrons: hydrogen-2 (deuterium), lithium-6, boron-10 and nitrogen-14. Also, only four naturally occurring, radioactive odd–odd nuclides have a half-life over a billion years: potassium-40, vanadium-50, lanthanum-138 and tantalum-180m. Most odd–odd nuclei are highly unstable with respect to beta decay, because the decay products are even–even, and are therefore more strongly bound, due to nuclear pairing effects.

Mass

The large majority of an atom's mass comes from the protons and neutrons that make it up. The total number of these particles (called "nucleons") in a given atom is called the mass number. It is a positive integer and dimensionless (instead of having dimension of mass), because it expresses a count. An example of use of a mass number is "carbon-12," which has 12 nucleons (six protons and six neutrons).

The actual mass of an atom at rest is often expressed using the unified atomic mass unit (u), also called dalton (Da). This unit is defined as a twelfth of the mass of a free neutral atom of carbon-12, which is approximately 1.66×10^{-27} kg. Hydrogen-1 (the lightest isotope of hydrogen which is also the nuclide with the lowest mass) has an atomic weight of 1.007825 u. The value of this number is called the atomic mass. A given atom has an atomic mass approximately equal (within 1%) to its mass number times the atomic mass unit (for example the mass of a nitrogen-14 is roughly 14 u). However, this number will not be exactly an integer except in the case of carbon-12. The heaviest stable atom is lead-208, with a mass of 207.9766521 u.

As even the most massive atoms are far too light to work with directly, chemists instead use the unit of moles. One mole of atoms of any element always has the same number of atoms (about 6.022×10^{23}). This number was chosen so that if an element has an atomic mass of 1 u, a mole of atoms of that element has a mass close to one gram. Because of the definition of the unified atomic mass unit, each carbon-12 atom has an atomic mass of exactly 12 u, and so a mole of carbon-12 atoms weighs exactly 0.012 kg.

Shape and Size

Atoms lack a well-defined outer boundary, so their dimensions are usually described in terms of an atomic radius. This is a measure of the distance out to which the electron cloud extends from the nucleus. However, this assumes the atom to exhibit a spherical shape, which is only obeyed for atoms in vacuum or free space. Atomic radii may be derived from the distances between two nuclei when the two atoms are joined in a chemical bond. The radius varies with the location of an atom on the atomic chart, the type of chemical bond, the number of neighboring atoms (coordination number) and a quantum mechanical property known as spin. On the periodic table of the elements, atom size tends to increase when moving down columns, but decrease when moving across rows (left to right). Consequently, the smallest atom is helium with a radius of 32 pm, while one of the largest is caesium at 225 pm.

When subjected to external forces, like electrical fields, the shape of an atom may deviate from spherical symmetry. The deformation depends on the field magnitude and the orbital type of outer shell electrons, as shown by group-theoretical considerations. Aspherical deviations might be elicited for instance in crystals, where large crystal-electrical fields may occur at low-symmetry lattice sites. Significant ellipsoidal deformations have recently been shown to occur for sulfur ions and chalcogen ions in pyrite-type compounds.

Atomic dimensions are thousands of times smaller than the wavelengths of light (400–700 nm) so they cannot be viewed using an optical microscope. However, individual atoms can be observed using a scanning tunneling microscope. To visualize the minuteness of the atom, consider that a typical human hair is about 1 million carbon atoms in width. A single drop of water contains about 2 sextillion (2×10^{21}) atoms of oxygen, and twice the number of hydrogen atoms. A single carat diamond with a mass of 2×10^{-4} kg contains about 10 sextillion (10^{22}) atoms of carbon.[note 2] If an apple were magnified to the size of the Earth, then the atoms in the apple would be approximately the size of the original apple.

Radioactive Decay

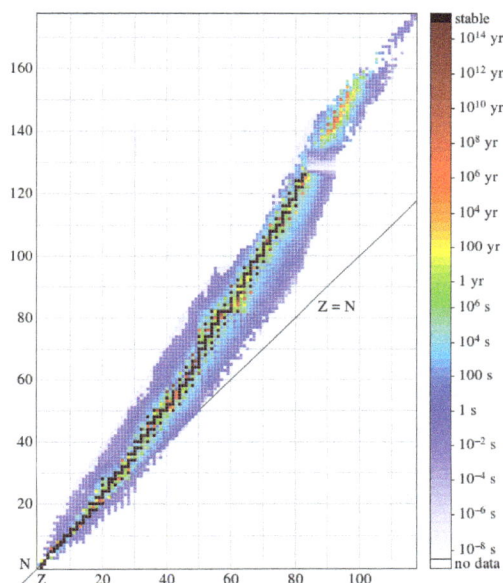

This diagram shows the half-life ($T_{1/2}$) of various isotopes with Z protons and N neutrons.

Every element has one or more isotopes that have unstable nuclei that are subject to radioactive decay, causing the nucleus to emit particles or electromagnetic radiation. Radioactivity can occur when the radius of a nucleus is large compared with the radius of the strong force, which only acts over distances on the order of 1 fm.

The most common forms of radioactive decay are:

- Alpha decay: this process is caused when the nucleus emits an alpha particle, which is a helium nucleus consisting of two protons and two neutrons. The result of the emission is a new element with a lower atomic number.

- Beta decay (and electron capture): these processes are regulated by the weak force, and result from a transformation of a neutron into a proton, or a proton into a neutron. The neutron to proton transition is accompanied by the emission of an electron and an antineutrino, while proton to neutron transition (except in electron capture) causes the emission of a positron and a neutrino. The electron or positron emissions are called beta particles. Beta decay either increases or decreases the atomic number of the nucleus by one. Electron capture is more common than positron emission, because it requires less energy. In this type of decay, an electron is absorbed by the nucleus, rather than a positron emitted from the nucleus. A neutrino is still emitted in this process, and a proton changes to a neutron.

- Gamma decay: this process results from a change in the energy level of the nucleus to a lower state, resulting in the emission of electromagnetic radiation. The excited state of a nucleus which results in gamma emission usually occurs following the emission of an alpha or a beta particle. Thus, gamma decay usually follows alpha or beta decay.

Other more rare types of radioactive decay include ejection of neutrons or protons or clusters of nucleons from a nucleus, or more than one beta particle. An analog of gamma emission which allows excited nuclei to lose energy in a different way, is internal conversion— a process that produces high-speed electrons that are not beta rays, followed by production of high-energy photons that are not gamma rays. A few large nuclei explode into two or more charged fragments of varying masses plus several neutrons, in a decay called spontaneous nuclear fission.

Each radioactive isotope has a characteristic decay time period—the half-life—that is determined by the amount of time needed for half of a sample to decay. This is an exponential decay process that steadily decreases the proportion of the remaining isotope by 50% every half-life. Hence after two half-lives have passed only 25% of the isotope is present, and so forth.

Magnetic Moment

Elementary particles possess an intrinsic quantum mechanical property known as spin. This is analogous to the angular momentum of an object that is spinning around its center of mass, although strictly speaking these particles are believed to be point-like and cannot be said to be rotating. Spin is measured in units of the reduced Planck constant ($\hbar$), with electrons, protons and neutrons all having spin $\frac{1}{2}\hbar$, or «spin-$\frac{1}{2}$». In an atom, electrons in motion around the nucleus possess orbital angular momentum in addition to their spin, while the nucleus itself possesses angular momentum due to its nuclear spin.

The magnetic field produced by an atom—its magnetic moment—is determined by these various forms of angular momentum, just as a rotating charged object classically produces a magnetic field. However, the most dominant contribution comes from electron spin. Due to the nature of electrons to obey the Pauli exclusion principle, in which no two electrons may be found in the same quantum state, bound electrons pair up with each other, with one member of each pair in a spin up state and the other in the opposite, spin down state. Thus these spins cancel each other out, reducing the total magnetic dipole moment to zero in some atoms with even number of electrons.

In ferromagnetic elements such as iron, cobalt and nickel, an odd number of electrons leads to an unpaired electron and a net overall magnetic moment. The orbitals of neighboring atoms overlap and a lower energy state is achieved when the spins of unpaired electrons are aligned with each other, a spontaneous process known as an exchange interaction. When the magnetic moments of ferromagnetic atoms are lined up, the material can produce a measurable macroscopic field. Paramagnetic materials have atoms with magnetic moments that line up in random directions when no magnetic field is present, but the magnetic moments of the individual atoms line up in the presence of a field.

The nucleus of an atom will have no spin when it has even numbers of both neutrons and protons, but for other cases of odd numbers, the nucleus may have a spin. Normally nuclei with spin are aligned in random directions because of thermal equilibrium. However, for certain elements (such as xenon-129) it is possible to polarize a significant proportion of the nuclear spin states so that they are aligned in the same direction—a condition called hyperpolarization. This has important applications in magnetic resonance imaging.

Energy Levels

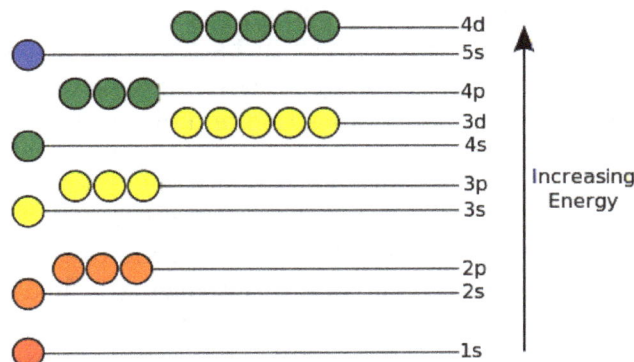

These electron's energy levels (not to scale) are sufficient for ground states of atoms up to cadmium ($5s^2$ $4d^{10}$) inclusively. Do not forget that even the top of the diagram is lower than an unbound electron state.

The potential energy of an electron in an atom is negative, its dependence of its position reaches the minimum (the most absolute value) inside the nucleus, and vanishes when the distance from the nucleus goes to infinity, roughly in an inverse proportion to the distance. In the quantum-mechanical model, a bound electron can only occupy a set of states centered on the nucleus, and each state corresponds to a specific energy level; see time-independent Schrödinger equation for theoretical explanation. An energy level can be measured by the amount of energy needed to unbind the electron from the atom, and is usually given in units of electronvolts (eV). The lowest energy state of a bound electron is called the ground state, i.e. stationary state, while an electron transition to a higher level results in an excited state. The electron's energy raises when n increases because the (average) distance to the nucleus increases. Dependence of the energy on ℓ is caused not by electrostatic potential of the nucleus, but by interaction between electrons.

For an electron to transition between two different states, e.g. grounded state to first excited level (ionization), it must absorb or emit a photon at an energy matching the difference in the potential energy of those levels, according to Niels Bohr model, what can be precisely calculated by the Schrödinger equation. Electrons jump between orbitals in a particle-like fashion. For example, if a single photon strikes the electrons, only a single electron changes states in response to the photon.

The energy of an emitted photon is proportional to its frequency, so these specific energy levels appear as distinct bands in the electromagnetic spectrum. Each element has a characteristic spectrum that can depend on the nuclear charge, subshells filled by electrons, the electromagnetic interactions between the electrons and other factors.

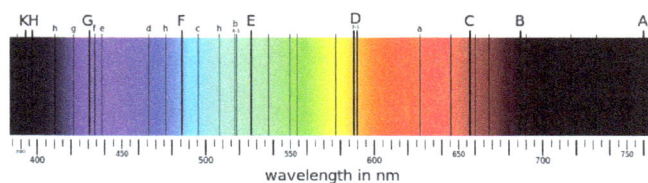

An example of absorption lines in a spectrum

When a continuous spectrum of energy is passed through a gas or plasma, some of the photons are absorbed by atoms, causing electrons to change their energy level. Those excited electrons that remain bound to their atom spontaneously emit this energy as a photon, traveling in a random direction, and so drop back to lower energy levels. Thus the atoms behave like a filter that forms a series of dark absorption bands in the energy output. (An observer viewing the atoms from a view that does not include the continuous spectrum in the background, instead sees a series of emission lines from the photons emitted by the atoms.) Spectroscopic measurements of the strength and width of atomic spectral lines allow the composition and physical properties of a substance to be determined.

Close examination of the spectral lines reveals that some display a fine structure splitting. This occurs because of spin–orbit coupling, which is an interaction between the spin and motion of the outermost electron. When an atom is in an external magnetic field, spectral lines become split into three or more components; a phenomenon called the Zeeman effect. This is caused by the interaction of the magnetic field with the magnetic moment of the atom and its electrons. Some atoms can have multiple electron configurations with the same energy level, which thus appear as a single spectral line. The interaction of the magnetic field with the atom shifts these electron configurations to slightly different energy levels, resulting in multiple spectral lines. The presence of an external electric field can cause a comparable splitting and shifting of spectral lines by modifying the electron energy levels, a phenomenon called the Stark effect.

If a bound electron is in an excited state, an interacting photon with the proper energy can cause stimulated emission of a photon with a matching energy level. For this to occur, the electron must drop to a lower energy state that has an energy difference matching the energy of the interacting photon. The emitted photon and the interacting photon then move off in parallel and with matching phases. That is, the wave patterns of the two photons are synchronized. This physical property is used to make lasers, which can emit a coherent beam of light energy in a narrow frequency band.

Valence and Bonding Behavior

Valency is the combining power of an element. It is equal to number of hydrogen atoms that atom can combine or displace in forming compounds. The outermost electron shell of an atom in its uncombined state is known as the valence shell, and the electrons in that shell are called valence electrons. The number of valence electrons determines the bonding behavior with other atoms. Atoms tend to chemically react with each other in a manner that fills (or empties) their outer valence shells. For example, a transfer of a single electron between atoms is a useful approximation for bonds that form between atoms with one-electron more than a filled shell, and others that are one-electron short of a full shell, such as occurs in the compound sodium chloride and other chemical ionic salts. However, many elements display multiple valences, or tendencies to share differing numbers of electrons in different compounds. Thus, chemical bonding between these elements takes many forms of electron-sharing that are more than simple electron transfers. Examples include the element carbon and the organic compounds.

The chemical elements are often displayed in a periodic table that is laid out to display recurring chemical properties, and elements with the same number of valence electrons form a group that is aligned in the same column of the table. (The horizontal rows correspond to the filling of a quantum shell of electrons.) The elements at the far right of the table have their outer shell completely filled with electrons, which results in chemically inert elements known as the noble gases.

States

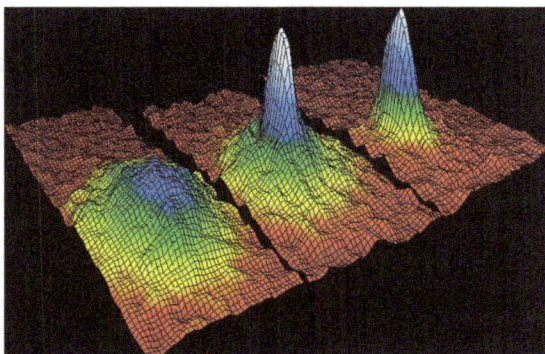

Snapshots illustrating the formation of a Bose–Einstein condensate

Quantities of atoms are found in different states of matter that depend on the physical conditions, such as temperature and pressure. By varying the conditions, materials can transition between solids, liquids, gases and plasmas. Within a state, a material can also exist in different allotropes. An example of this is solid carbon, which can exist as graphite or diamond. Gaseous allotropes exist as well, such as dioxygen and ozone.

At temperatures close to absolute zero, atoms can form a Bose–Einstein condensate, at which point quantum mechanical effects, which are normally only observed at the atomic scale, become apparent on a macroscopic scale. This super-cooled collection of atoms then behaves as a single super atom, which may allow fundamental checks of quantum mechanical behavior.

Identification

Scanning tunneling microscope image showing the individual atoms making up this gold (100) surface. The surface atoms deviate from the bulk crystal structure and arrange in columns several atoms wide with pits between them.

The scanning tunneling microscope is a device for viewing surfaces at the atomic level. It uses the quantum tunneling phenomenon, which allows particles to pass through a barrier that would normally be insurmountable. Electrons tunnel through the vacuum between two planar metal electrodes, on each of which is an adsorbed atom, providing a tunneling-current density that can be measured. Scanning one atom (taken as the tip) as it moves past the other (the sample) permits plotting of tip displacement versus lateral separation for a constant current. The calculation shows the extent to which scanning-tunneling-microscope images of an individual atom are visible. It confirms that for low bias, the microscope images the space-averaged dimensions of the electron orbitals across closely packed energy levels—the Fermi level local density of states.

An atom can be ionized by removing one of its electrons. The electric charge causes the trajectory of an atom to bend when it passes through a magnetic field. The radius by which the trajectory of a moving ion is turned by the magnetic field is determined by the mass of the atom. The mass spectrometer uses this principle to measure the mass-to-charge ratio of ions. If a sample contains multiple isotopes, the mass spectrometer can determine the proportion of each isotope in the sample by measuring the intensity of the different beams of ions. Techniques to vaporize atoms include inductively coupled plasma atomic emission spectroscopy and inductively coupled plasma mass spectrometry, both of which use a plasma to vaporize samples for analysis.

A more area-selective method is electron energy loss spectroscopy, which measures the energy loss of an electron beam within a transmission electron microscope when it interacts with a portion of a sample. The atom-probe tomograph has sub-nanometer resolution in 3-D and can chemically identify individual atoms using time-of-flight mass spectrometry.

Spectra of excited states can be used to analyze the atomic composition of distant stars. Specific light wavelengths contained in the observed light from stars can be separated out and related to the quantized transitions in free gas atoms. These colors can be replicated using a gas-discharge lamp containing the same element. Helium was discovered in this way in the spectrum of the Sun 23 years before it was found on Earth.

Origin and Current State

Atoms form about 4% of the total energy density of the observable Universe, with an average density of about 0.25 atoms/m^3. Within a galaxy such as the Milky Way, atoms have a much higher concentration, with the density of matter in the interstellar medium (ISM) ranging from 10^5 to 10^9 atoms/m^3. The Sun is believed to be inside the Local Bubble, a region of highly ionized gas, so the density in the solar neighborhood is only about 10^3 atoms/m^3. Stars form from dense clouds in the ISM, and the evolutionary processes of stars result in the steady enrichment of the ISM with elements more massive than hydrogen and helium. Up to 95% of the Milky Way's atoms are concentrated inside stars and the total mass of atoms forms about 10% of the mass of the galaxy. (The remainder of the mass is an unknown dark matter.)

Formation

Electrons are thought to exist in the Universe since early stages of the Big Bang. Atomic nuclei forms in nucleosynthesis reactions. In about three minutes Big Bang nucleosynthesis produced most of the helium, lithium, and deuterium in the Universe, and perhaps some of the beryllium and boron.

Ubiquitousness and stability of atoms relies on their binding energy, which means that an atom has a lower energy than an unbound system of the nucleus and electrons. Where the temperature is much higher than ionization potential, the matter exists in the form of plasma—a gas of positively charged ions (possibly, bare nuclei) and electrons. When the temperature drops below the ionization potential, atoms become statistically favorable. Atoms (complete with bound electrons) became to dominate over charged particles 380,000 years after the Big Bang—an epoch called recombination, when the expanding Universe cooled enough to allow electrons to become attached to nuclei.

Since the Big Bang, which produced no carbon or heavier elements, atomic nuclei have been combined in stars through the process of nuclear fusion to produce more of the element helium, and (via the triple alpha process) the sequence of elements from carbon up to iron.

Isotopes such as lithium-6, as well as some beryllium and boron are generated in space through cosmic ray spallation. This occurs when a high-energy proton strikes an atomic nucleus, causing large numbers of nucleons to be ejected.

Elements heavier than iron were produced in supernovae through the r-process and in AGB stars through the s-process, both of which involve the capture of neutrons by atomic nuclei. Elements such as lead formed largely through the radioactive decay of heavier elements.

Earth

Most of the atoms that make up the Earth and its inhabitants were present in their current form in the nebula that collapsed out of a molecular cloud to form the Solar System. The rest are the result of radioactive decay, and their relative proportion can be used to determine the age of the Earth through radiometric dating. Most of the helium in the crust of the Earth (about 99% of the helium from gas wells, as shown by its lower abundance of helium-3) is a product of alpha decay.

There are a few trace atoms on Earth that were not present at the beginning (i.e., not "primordial"), nor are results of radioactive decay. Carbon-14 is continuously generated by cosmic rays in the atmosphere. Some atoms on Earth have been artificially generated either deliberately or as by-products of nuclear reactors or explosions. Of the transuranic elements—those with atomic numbers greater than 92—only plutonium and neptunium occur naturally on Earth. Transuranic elements have radioactive lifetimes shorter than the current age of the Earth and thus identifiable quantities of these elements have long since decayed, with the exception of traces of plutonium-244 possibly deposited by cosmic dust. Natural deposits of plutonium and neptunium are produced by neutron capture in uranium ore.

The Earth contains approximately 1.33×10^{50} atoms. Although small numbers of independent atoms of noble gases exist, such as argon, neon, and helium, 99% of the atmosphere is bound in the form of molecules, including carbon dioxide and diatomic oxygen and nitrogen. At the surface of the Earth, an overwhelming majority of atoms combine to form various compounds, including water, salt, silicates and oxides. Atoms can also combine to create materials that do not consist of discrete molecules, including crystals and liquid or solid metals. This atomic matter forms networked arrangements that lack the particular type of small-scale interrupted order associated with molecular matter.

Rare and Theoretical Forms

Superheavy Elements

While isotopes with atomic numbers higher than lead (82) are known to be radioactive, an "island of stability" has been proposed for some elements with atomic numbers above 103. These super-heavy elements may have a nucleus that is relatively stable against radioactive decay. The most likely candidate for a stable superheavy atom, unbihexium, has 126 protons and 184 neutrons.

Exotic Matter

Each particle of matter has a corresponding antimatter particle with the opposite electrical charge. Thus, the positron is a positively charged antielectron and the antiproton is a negatively charged equivalent of a proton. When a matter and corresponding antimatter particle meet, they annihilate each other. Because of this, along with an imbalance between the number of matter and antimatter particles, the latter are rare in the universe. The first causes of this imbalance are not yet fully understood, although theories of baryogenesis may offer an explanation. As a result, no antimatter atoms have been discovered in nature. However, in 1996 the antimatter counterpart of the hydrogen atom (antihydrogen) was synthesized at the CERN laboratory in Geneva.

Other exotic atoms have been created by replacing one of the protons, neutrons or electrons with other particles that have the same charge. For example, an electron can be replaced by a more massive muon, forming a muonic atom. These types of atoms can be used to test the fundamental predictions of physics.

Components of Atom

Ion

An ion is an electrically charged atom or group of atoms. It is a part of an atom, or part of a group of atoms (molecule). It is "charged" so it will move near electricity. This is because atoms are made of three smaller parts (1) neutrons (with no charge), and equal numbers of (2) charged protons and (3) oppositely-charged electrons. An ion has unequal numbers of protons and electrons. Making an ion from an atom or molecule is called *ionization*.

The charge on a proton is measured as +1 (positively charged), and the charge on an electron is measured as -1 (negatively charged). An atom that is ionized makes two ions, one positive, and one negatively charged. For example, a neutral hydrogen atom has one proton and one electron. Heating the atom breaks it into two parts (1) a positively charged hydrogen ion, H+ (2) a negatively charged electron.

A liquid with ions is called an electrolyte. A gas with lots of ions is called a plasma. When ions move, it is called electricity. For example, in a wire, the metal ions do not move, but the electrons move as electricity. A positive ion and a negative ion will move together. Two ions of the same charge will move apart. When ions move they also make *magnetic fields*.

Chemistry

In chemistry, a molecule or atom that is electrically charged is called an ion. An ion has more or less electrons than there are protons in an atom. The process of giving or taking electrons away from a normal atom and turning it into an ion is called ionization.

Many ions are colourless. Elements in the main groups in the Periodic Table form colourless ions. Some ions are coloured. The transition metals usually form coloured ions.

Physics

In physics, atomic nuclei that have been completely ionized are called *charged particles*. These are ones in alpha radiation.

Ionization happens by giving atoms high energy. This is done using electrical voltage or by high-energy ionizing radiation or high temperature.

An ionized gas is called plasma.

A simple ion is formed from a single atom.

Polyatomic ions are formed from a lot of atoms. Polyatomic ions usually consist of all non-metal atoms, but sometimes the polyatomic ion can have a metallic atom too.

Positive ions are called cations. They are attracted to cathodes (negatively charged electrodes). (Cation is pronounced "cat eye on", not "kay shun".) All simple metal ions are cations.

Negative ions are called **anions**. They are attracted to anodes (positively charged electrodes). All simple non-metal ions (except H^+, which is a proton) are anions (except NH_4^+).

Transition metals can form more than one simple cation with different charges.

Most ions have a charge of less than 4, but some can have higher charges.

Michael Faraday was the first person to write a theory about ions, in 1830. In his theory, he said what the portions of molecules were like that moved to anions or cations. Svante August Arrhenius showed how this happened. He wrote this in his doctoral dissertation in 1884 (University of Uppsala). The university did not accept his theory at first (he only just passed his degree). But in 1903, he won the Nobel Prize in Chemistry for the same idea.

In Greek *ion* is like the word "go". "Anion" and "cation" mean "up-goer" and "down-goer". "Anode" and "cathode" are "way up" and "way down".

Other Meanings

In Greek mythology, **Ion** was a son of Xuthus and Creusa. He founded the Ionian race and became a king of Athens. The term is also used for an element of the Plato texts, and a Window manager.

Electron

The **electron** is a subatomic particle, symbol e– or β–, with a negative elementary electric charge. Electrons belong to the first generation of the lepton particle family, and are generally thought to be elementary particles because they have no known components or substructure. The electron has a mass that is approximately 1/1836 that of the proton. Quantum mechanical properties of the electron include an intrinsic angular momentum (spin) of a half-integer value in units of $\hbar$, which means that it is a fermion. Being fermions, no two electrons can occupy the same quantum state, in accordance with the Pauli exclusion principle. Like all matter, electrons have properties of both particles and waves, and so can collide with other particles and can be diffracted like light. The wave properties of electrons are easier to observe with experiments than those of other particles like neutrons and protons because electrons have a lower mass and hence a higher De Broglie wavelength for typical energies.

Many physical phenomena involve electrons in an essential role, such as electricity, magnetism, and thermal conductivity, and they also participate in gravitational, electromagnetic and weak interactions. An electron generates an electric field surrounding it. An electron moving relative to an observer generates a magnetic field. External electromagnetic fields affect an electron according to the Lorentz force law. Electrons radiate or absorb energy in the form of photons when accelerated. Laboratory instruments are capable of containing and observing individual electrons as well as electron plasma using electromagnetic fields, whereas dedicated telescopes can detect electron plasma in outer space. Electrons are involved in many applications such as electronics, welding, cathode ray tubes, electron microscopes, radiation therapy, lasers, gaseous ionization detectors and particle accelerators.

Interactions involving electrons and other subatomic particles are of interest in fields such as chemistry and nuclear physics. The Coulomb force interaction between positive protons inside atomic nuclei and negative electrons composes atoms. Ionization or changes in the proportions of particles changes the binding energy of the system. The exchange or sharing of the electrons between two or more atoms is the main cause of chemical bonding. British natural philosopher Richard Laming first hypothesized the concept of an indivisible quantity of electric charge to explain the chemical properties of atoms in 1838; Irish physicist George Johnstone Stoney named this charge 'electron' in 1891, and J. J. Thomson and his team of British physicists identified it as a particle in 1897. Electrons can also participate in nuclear reactions, such as nucleosynthesis in stars, where they are known as beta particles. Electrons may be created through beta decay of radioactive isotopes and in high-energy collisions, for instance when cosmic rays enter the atmosphere. The antiparticle of the electron is called the positron; it is identical to the electron except that it carries electrical and other charges of the opposite sign. When an electron collides with a positron, both particles may be totally annihilated, producing gamma ray photons.

History

The ancient Greeks noticed that amber attracted small objects when rubbed with fur. Along with lightning, this phenomenon is one of humanity's earliest recorded experiences with electricity. In his 1600 treatise *De Magnete*, the English scientist William Gilbert coined the New Latin term *electricus*, to refer to this property of attracting small objects after being rubbed. Both *electric* and *electricity* are derived from the Latin *ēlectrum* (also the root of the alloy of the same name), which came from the Greek word for amber, ἤλεκτρον (*ēlektron*).

In the early 1700s, Francis Hauksbee and French chemist Charles François de Fay independently discovered what they believed were two kinds of frictional electricity—one generated from rubbing glass, the other from rubbing resin. From this, Du Fay theorized that electricity consists of two electrical fluids, *vitreous* and *resinous*, that are separated by friction, and that neutralize each other when combined. A decade later Benjamin Franklin proposed that electricity was not from different types of electrical fluid, but the same electrical fluid under different pressures. He gave them the modern charge nomenclature of positive and negative respectively. Franklin thought of the charge carrier as being positive, but he did not correctly identify which situation was a surplus of the charge carrier, and which situation was a deficit.

Between 1838 and 1851, British natural philosopher Richard Laming developed the idea that an atom is composed of a core of matter surrounded by subatomic particles that had unit electric charges. Beginning in 1846, German physicist William Weber theorized that electricity was composed of positively and negatively charged fluids, and their interaction was governed by the inverse square law. After studying the phenomenon of electrolysis in 1874, Irish physicist George Johnstone Stoney suggested that there existed a "single definite quantity of electricity", the charge of a monovalent ion. He was able to estimate the value of this elementary charge *e* by means of Faraday's laws of electrolysis. However, Stoney believed these charges were permanently attached to atoms and could not be removed. In 1881, German physicist Hermann von Helmholtz argued that both positive and negative charges were divided into elementary parts, each of which "behaves like atoms of electricity".

Stoney initially coined the term *electrolion* in 1881. Ten years later, he switched to *electron* to describe these elementary charges, writing in 1894: "... an estimate was made of the actual amount of this most remarkable fundamental unit of electricity, for which I have since ventured to suggest the name *electron*". A 1906 proposal to change to *electrion* failed because Hendrik Lorentz preferred to keep *electron*. The word *electron* is a combination of the words *electric* and *ion*. The suffix *-on* which is now used to designate other subatomic particles, such as a proton or neutron, is in turn derived from electron.

Discovery

A beam of electrons deflected in a circle by a magnetic field

The German physicist Johann Wilhelm Hittorf studied electrical conductivity in rarefied gases: in 1869, he discovered a glow emitted from the cathode that increased in size with decrease in gas pressure. In 1876, the German physicist Eugen Goldstein showed that the rays from this glow cast a shadow, and he dubbed the rays cathode rays. During the 1870s, the English chemist and physicist Sir William Crookes developed the first cathode ray tube to have a high vacuum inside.

He then showed that the luminescence rays appearing within the tube carried energy and moved from the cathode to the anode. Furthermore, by applying a magnetic field, he was able to deflect the rays, thereby demonstrating that the beam behaved as though it were negatively charged. In 1879, he proposed that these properties could be explained by what he termed 'radiant matter'. He suggested that this was a fourth state of matter, consisting of negatively charged molecules that were being projected with high velocity from the cathode.

The German-born British physicist Arthur Schuster expanded upon Crookes' experiments by placing metal plates parallel to the cathode rays and applying an electric potential between the plates. The field deflected the rays toward the positively charged plate, providing further evidence that the rays carried negative charge. By measuring the amount of deflection for a given level of current, in 1890 Schuster was able to estimate the charge-to-mass ratio of the ray components. However, this produced a value that was more than a thousand times greater than what was expected, so little credence was given to his calculations at the time.

In 1892 Hendrik Lorentz suggested that the mass of these particles (electrons) could be a consequence of their electric charge.

In 1896, the British physicist J. J. Thomson, with his colleagues John S. Townsend and H. A. Wilson, performed experiments indicating that cathode rays really were unique particles, rather than waves, atoms or molecules as was believed earlier. Thomson made good estimates of both the charge e and the mass m, finding that cathode ray particles, which he called "corpuscles," had perhaps one thousandth of the mass of the least massive ion known: hydrogen. He showed that their charge to mass ratio, e/m, was independent of cathode material. He further showed that the negatively charged particles produced by radioactive materials, by heated materials and by illuminated materials were universal. The name electron was again proposed for these particles by the Irish physicist George F. Fitzgerald, and the name has since gained universal acceptance.

Robert Millikan

While studying naturally fluorescing minerals in 1896, the French physicist Henri Becquerel discovered that they emitted radiation without any exposure to an external energy source. These radioactive materials became the subject of much interest by scientists, including the New Zealand physicist Ernest Rutherford who discovered they emitted particles. He designated these particles alpha and beta, on the basis of their ability to penetrate matter. In 1900, Becquerel showed that the beta rays emitted by radium could be deflected by an electric field, and that their mass-to-charge ratio was the same as for cathode rays. This evidence strengthened the view that electrons existed as components of atoms.

The electron's charge was more carefully measured by the American physicists Robert Millikan and Harvey Fletcher in their oil-drop experiment of 1909, the results of which were published in 1911. This experiment used an electric field to prevent a charged droplet of oil from falling as a result of gravity. This device could measure the electric charge from as few as 1–150 ions with an error margin of less than 0.3%. Comparable experiments had been done earlier by Thomson's team, using clouds of charged water droplets generated by electrolysis, and in 1911 by Abram Ioffe, who independently obtained the same result as Millikan using charged microparticles of metals, then published his results in 1913. However, oil drops were more stable than water drops because of their slower evaporation rate, and thus more suited to precise experimentation over longer periods of time.

Around the beginning of the twentieth century, it was found that under certain conditions a fast-moving charged particle caused a condensation of supersaturated water vapor along its path. In 1911, Charles Wilson used this principle to devise his cloud chamber so he could photograph the tracks of charged particles, such as fast-moving electrons.

Atomic Theory

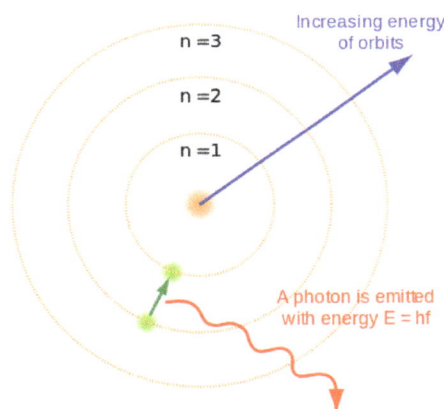

The Bohr model of the atom, showing states of electron with energy quantized by the number n. An electron dropping to a lower orbit emits a photon equal to the energy difference between the orbits.

By 1914, experiments by physicists Ernest Rutherford, Henry Moseley, James Franck and Gustav Hertz had largely established the structure of an atom as a dense nucleus of positive charge surrounded by lower-mass electrons. In 1913, Danish physicist Niels Bohr postulated that electrons resided in quantized energy states, with the energy determined by the angular momentum of the electron's orbits about the nucleus. The electrons could move between these states, or orbits, by the emission or absorption of photons at specific frequencies. By means of these quantized orbits, he accurately explained the spectral lines of the hydrogen atom. However, Bohr's model failed to account for the relative intensities of the spectral lines and it was unsuccessful in explaining the spectra of more complex atoms.

Chemical bonds between atoms were explained by Gilbert Newton Lewis, who in 1916 proposed that a covalent bond between two atoms is maintained by a pair of electrons shared between them.

Later, in 1927, Walter Heitler and Fritz London gave the full explanation of the electron-pair formation and chemical bonding in terms of quantum mechanics. In 1919, the American chemist Irving Langmuir elaborated on the Lewis' static model of the atom and suggested that all electrons were distributed in successive "concentric (nearly) spherical shells, all of equal thickness". The shells were, in turn, divided by him in a number of cells each containing one pair of electrons. With this model Langmuir was able to qualitatively explain the chemical properties of all elements in the periodic table, which were known to largely repeat themselves according to the periodic law.

In 1924, Austrian physicist Wolfgang Pauli observed that the shell-like structure of the atom could be explained by a set of four parameters that defined every quantum energy state, as long as each state was inhabited by no more than a single electron. (This prohibition against more than one electron occupying the same quantum energy state became known as the Pauli exclusion principle.) The physical mechanism to explain the fourth parameter, which had two distinct possible values, was provided by the Dutch physicists Samuel Goudsmit and George Uhlenbeck. In 1925, Goudsmit and Uhlenbeck suggested that an electron, in addition to the angular momentum of its orbit, possesses an intrinsic angular momentum and magnetic dipole moment. The intrinsic angular momentum became known as spin, and explained the previously mysterious splitting of spectral lines observed with a high-resolution spectrograph; this phenomenon is known as fine structure splitting.

Quantum Mechanics

In his 1924 dissertation *Recherches sur la théorie des quanta* (Research on Quantum Theory), French physicist Louis de Broglie hypothesized that all matter possesses a de Broglie wave similar to light. That is, under the appropriate conditions, electrons and other matter would show properties of either particles or waves. The corpuscular properties of a particle are demonstrated when it is shown to have a localized position in space along its trajectory at any given moment. Wave-like nature is observed, for example, when a beam of light is passed through parallel slits and creates interference patterns. In 1927, the interference effect was found in a beam of electrons by English physicist George Paget Thomson with a thin metal film and by American physicists Clinton Davisson and Lester Germer using a crystal of nickel.

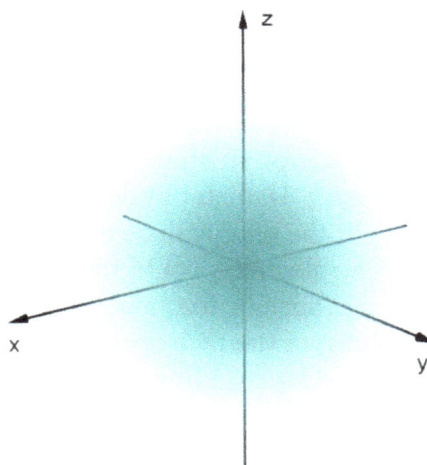

Orbital s ($\ell = 0$, $m_\ell = 0$)

In quantum mechanics, the behavior of an electron in an atom is described by an orbital, which is a probability distribution rather than an orbit. In the figure, the shading indicates the relative probability to "find" the electron, having the energy corresponding to the given quantum numbers, at that point.

De Broglie's prediction of a wave nature for electrons led Erwin Schrödinger to postulate a wave equation for electrons moving under the influence of the nucleus in the atom. In 1926, this equation, the Schrödinger equation, successfully described how electron waves propagated. Rather than yielding a solution that determined the location of an electron over time, this wave equation also could be used to predict the probability of finding an electron near a position, especially a position near where the electron was bound in space, for which the electron wave equations did not change in time. This approach led to a second formulation of quantum mechanics (the first being by Heisenberg in 1925), and solutions of Schrödinger's equation, like Heisenberg's, provided derivations of the energy states of an electron in a hydrogen atom that were equivalent to those that had been derived first by Bohr in 1913, and that were known to reproduce the hydrogen spectrum. Once spin and the interaction between multiple electrons were considered, quantum mechanics later made it possible to predict the configuration of electrons in atoms with higher atomic numbers than hydrogen.

In 1928, building on Wolfgang Pauli's work, Paul Dirac produced a model of the electron – the Dirac equation, consistent with relativity theory, by applying relativistic and symmetry considerations to the hamiltonian formulation of the quantum mechanics of the electro-magnetic field. To resolve some problems within his relativistic equation, in 1930 Dirac developed a model of the vacuum as an infinite sea of particles having negative energy, which was dubbed the Dirac sea. This led him to predict the existence of a positron, the antimatter counterpart of the electron. This particle was discovered in 1932 by Carl Anderson, who proposed calling standard electrons *negatrons*, and using *electron* as a generic term to describe both the positively and negatively charged variants.

In 1947 Willis Lamb, working in collaboration with graduate student Robert Retherford, found that certain quantum states of hydrogen atom, which should have the same energy, were shifted in relation to each other, the difference being the Lamb shift. About the same time, Polykarp Kusch, working with Henry M. Foley, discovered the magnetic moment of the electron is slightly larger than predicted by Dirac's theory. This small difference was later called anomalous magnetic dipole moment of the electron. This difference was later explained by the theory of quantum electrodynamics, developed by Sin-Itiro Tomonaga, Julian Schwinger and Richard Feynman in the late 1940s.

Particle Accelerators

With the development of the particle accelerator during the first half of the twentieth century, physicists began to delve deeper into the properties of subatomic particles. The first successful attempt to accelerate electrons using electromagnetic induction was made in 1942 by Donald Kerst. His initial betatron reached energies of 2.3 MeV, while subsequent betatrons achieved 300 MeV. In 1947, synchrotron radiation was discovered with a 70 MeV electron synchrotron at General Electric. This radiation was caused by the acceleration of electrons, moving near the speed of light, through a magnetic field.

With a beam energy of 1.5 GeV, the first high-energy particle collider was ADONE, which began operations in 1968. This device accelerated electrons and positrons in opposite directions, effectively doubling the energy of their collision when compared to striking a static target with an electron. The Large Electron–Positron Collider (LEP) at CERN, which was operational from 1989 to 2000, achieved collision energies of 209 GeV and made important measurements for the Standard Model of particle physics.

Confinement of Individual Electrons

Individual electrons can now be easily confined in ultra small (L = 20 nm, W = 20 nm) CMOS transistors operated at cryogenic temperature over a range of −269 °C (4 K) to about −258 °C (15 K). The electron wavefunction spreads in a semiconductor lattice and negligibly interacts with the valence band electrons, so it can be treated in the single particle formalism, by replacing its mass with the effective mass tensor.

Characteristics

Classification

Standard Model of elementary particles. The electron (symbol e) is on the left.

In the Standard Model of particle physics, electrons belong to the group of subatomic particles called leptons, which are believed to be fundamental or elementary particles. Electrons have the lowest mass of any charged lepton (or electrically charged particle of any type) and belong to the first-generation of fundamental particles. The second and third generation contain charged leptons, the muon and the tau, which are identical to the electron in charge, spin and interactions, but are more massive. Leptons differ from the other basic constituent of matter, the quarks, by their lack of strong interaction. All members of the lepton group are fermions, because they all have half-odd integer spin; the electron has spin 1/2.

Fundamental Properties

The invariant mass of an electron is approximately 9.109×10^{-31} kilograms, or 5.489×10^{-4} atomic mass units. On the basis of Einstein's principle of mass–energy equivalence, this mass corresponds to a rest energy of 0.511 MeV. The ratio between the mass of a proton and that of an electron is about 1836. Astronomical measurements show that the proton-to-electron mass ratio has held the same value for at least half the age of the universe, as is predicted by the Standard Model.

Electrons have an electric charge of -1.602×10^{-19} coulomb, which is used as a standard unit of charge for subatomic particles, and is also called the elementary charge. This elementary charge has a relative standard uncertainty of 2.2×10^{-8}. Within the limits of experimental accuracy, the electron charge is identical to the charge of a proton, but with the opposite sign. As the symbol e is used for the elementary charge, the electron is commonly symbolized by e−, where the minus sign indicates the negative charge. The positron is symbolized by e+ because it has the same properties as the electron but with a positive rather than negative charge.

The electron has an intrinsic angular momentum or spin of 1/2. This property is usually stated by referring to the electron as a spin-1/2 particle. For such particles the spin magnitude is $\sqrt{3}/2\ \hbar$.[note 3] while the result of the measurement of a projection of the spin on any axis can only be $\pm\hbar/2$. In addition to spin, the electron has an intrinsic magnetic moment along its spin axis. It is approximately equal to one Bohr magneton,[note 4] which is a physical constant equal to $9.27400915(23)\times10^{-24}$ joules per tesla. The orientation of the spin with respect to the momentum of the electron defines the property of elementary particles known as helicity.

The electron has no known substructure. and it is assumed to be a point particle with a point charge and no spatial extent. In classical physics, the angular momentum and magnetic moment of an object depend upon its physical dimensions. Hence, the concept of a dimensionless electron possessing these properties contrasts to experimental observations in Penning traps which point to finite non-zero radius of the electron. A possible explanation of this paradoxical situation is given below in the "Virtual particles" subsection by taking into consideration the Foldy-Wouthuysen transformation.

The issue of the radius of the electron is a challenging problem of the modern theoretical physics. The admission of the hypothesis of a finite radius of the electron is incompatible to the premises of the theory of relativity. On the other hand, a point-like electron (zero radius) generates serious mathematical difficulties due to the self-energy of the electron tending to infinity. These aspects have been analyzed in detail by Dmitri Ivanenko and Arseny Sokolov.

Observation of a single electron in a Penning trap shows the upper limit of the particle's radius is 10^{-22} meters. Also an upper bound of electron radius of 10^{-18} meters can be derived using the uncertainty relation in energy.

There *is* also a physical constant called the "classical electron radius", with the much larger value of 2.8179×10^{-15} m, greater than the radius of the proton. However, the terminology comes from a simplistic calculation that ignores the effects of quantum mechanics; in reality, the so-called classical electron radius has little to do with the true fundamental structure of the electron.[note 5]

There are elementary particles that spontaneously decay into less massive particles. An example is the muon, which decays into an electron, a neutrino and an antineutrino, with a mean lifetime of 2.2×10^{-6} seconds. However, the electron is thought to be stable on theoretical grounds: the electron is the least massive particle with non-zero electric charge, so its decay would violate charge conservation. The experimental lower bound for the electron's mean lifetime is 6.6×10^{28} years, at a 90% confidence level.

Quantum Properties

As with all particles, electrons can act as waves. This is called the wave–particle duality and can be demonstrated using the double-slit experiment.

The wave-like nature of the electron allows it to pass through two parallel slits simultaneously, rather than just one slit as would be the case for a classical particle. In quantum mechanics, the wave-like property of one particle can be described mathematically as a complex-valued function, the wave function, commonly denoted by the Greek letter psi (ψ). When the absolute value of this function is squared, it gives the probability that a particle will be observed near a location—a probability density.

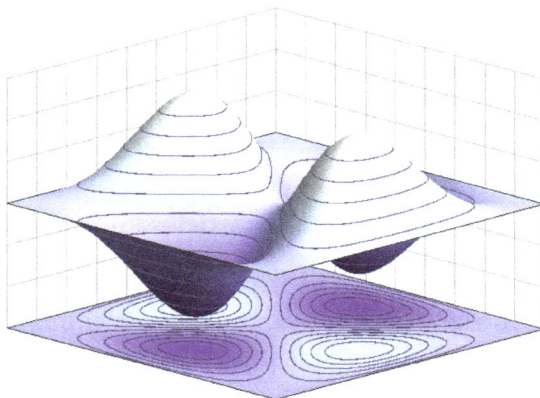

Example of an antisymmetric wave function for a quantum state of two identical fermions in a 1-dimensional box. If the particles swap position, the wave function inverts its sign.

Electrons are identical particles because they cannot be distinguished from each other by their intrinsic physical properties. In quantum mechanics, this means that a pair of interacting electrons must be able to swap positions without an observable change to the state of the system. The wave function of fermions, including electrons, is antisymmetric, meaning that it changes sign when two electrons are swapped; that is, $\psi(r_1, r_2) = -\psi(r_2, r_1)$, where the variables r_1 and r_2 correspond to the first and second electrons, respectively. Since the absolute value is not changed by a sign swap, this corresponds to equal probabilities. Bosons, such as the photon, have symmetric wave functions instead.

In the case of antisymmetry, solutions of the wave equation for interacting electrons result in a zero probability that each pair will occupy the same location or state. This is responsible for the Pauli exclusion principle, which precludes any two electrons from occupying the same quantum state. This principle explains many of the properties of electrons. For example, it causes groups of bound electrons to occupy different orbitals in an atom, rather than all overlapping each other in the same orbit.

Virtual Particles

In a simplified picture, every photon spends some time as a combination of a virtual electron plus its antiparticle, the virtual positron, which rapidly annihilate each other shortly thereafter. The combination of the energy variation needed to create these particles, and the time during which they exist, fall under the threshold of detectability expressed by the Heisenberg uncertainty relation, $\Delta E \cdot \Delta t \geq \hbar$. In effect, the energy needed to create these virtual particles, ΔE, can be "borrowed" from the vacuum for a period of time, Δt, so that their product is no more than the reduced Planck constant, $\hbar \approx 6.6 \times 10^{-16}$ eV·s. Thus, for a virtual electron, Δt is at most 1.3×10^{-21} s.

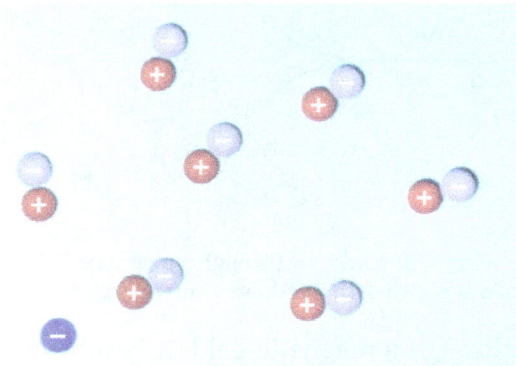

A schematic depiction of virtual electron–positron pairs appearing at random near an electron (at lower left)

While an electron–positron virtual pair is in existence, the coulomb force from the ambient electric field surrounding an electron causes a created positron to be attracted to the original electron, while a created electron experiences a repulsion. This causes what is called vacuum polarization. In effect, the vacuum behaves like a medium having a dielectric permittivity more than unity. Thus the effective charge of an electron is actually smaller than its true value, and the charge decreases with increasing distance from the electron. This polarization was confirmed experimentally in 1997 using the Japanese TRISTAN particle accelerator. Virtual particles cause a comparable shielding effect for the mass of the electron.

The interaction with virtual particles also explains the small (about 0.1%) deviation of the intrinsic magnetic moment of the electron from the Bohr magneton (the anomalous magnetic moment). The extraordinarily precise agreement of this predicted difference with the experimentally determined value is viewed as one of the great achievements of quantum electrodynamics.

The apparent paradox (mentioned above in the properties subsection) of a point particle electron having intrinsic angular momentum and magnetic moment can be explained by the formation of virtual photons in the electric field generated by the electron. These photons cause the electron to shift about in a jittery fashion (known as zitterbewegung), which results in a net circular motion with precession. This motion produces both the spin and the magnetic moment of the electron. In atoms, this creation of virtual photons explains the Lamb shift observed in spectral lines.

Interaction

An electron generates an electric field that exerts an attractive force on a particle with a positive charge, such as the proton, and a repulsive force on a particle with a negative charge. The strength of this force is determined by Coulomb's inverse square law. When an electron is in motion, it generates a magnetic field. The Ampère-Maxwell law relates the magnetic field to the mass motion of electrons (the current) with respect to an observer. This property of induction supplies the magnetic field that drives an electric motor. The electromagnetic field of an arbitrary moving charged particle is expressed by the Liénard–Wiechert potentials, which are valid even when the particle's speed is close to that of light (relativistic).

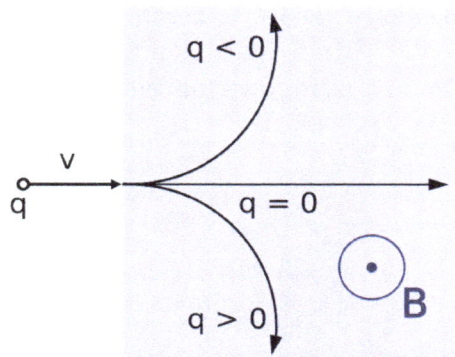

A particle with charge q (at left) is moving with velocity v through a magnetic field B that is oriented toward the viewer. For an electron, q is negative so it follows a curved trajectory toward the top.

When an electron is moving through a magnetic field, it is subject to the Lorentz force that acts perpendicularly to the plane defined by the magnetic field and the electron velocity. This centripetal force causes the electron to follow a helical trajectory through the field at a radius called the gyroradius. The acceleration from this curving motion induces the electron to radiate energy in the form of synchrotron radiation. The energy emission in turn causes a recoil of the electron, known as the Abraham–Lorentz–Dirac Force, which creates a friction that slows the electron. This force is caused by a back-reaction of the electron's own field upon itself.

Photons mediate electromagnetic interactions between particles in quantum electrodynamics. An isolated electron at a constant velocity cannot emit or absorb a real photon; doing so would violate conservation of energy and momentum. Instead, virtual photons can transfer momentum between two charged particles. This exchange of virtual photons, for example, generates the Coulomb force. Energy emission can occur when a moving electron is deflected by a charged particle, such as a proton. The acceleration of the electron results in the emission of Bremsstrahlung radiation.

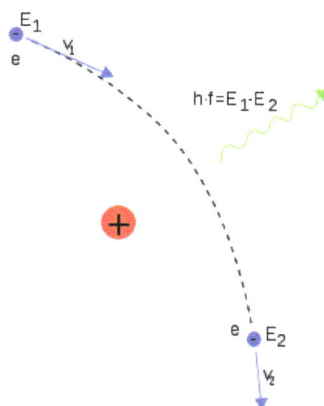

Here, Bremsstrahlung is produced by an electron e deflected by the electric field of an atomic nucleus. The energy change $E_2 - E_1$ determines the frequency f of the emitted photon.

An inelastic collision between a photon (light) and a solitary (free) electron is called Compton scattering. This collision results in a transfer of momentum and energy between the particles, which modifies the wavelength of the photon by an amount called the Compton shift.[note 7] The maximum magnitude of this wavelength shift is $h/m_e c$, which is known as the Compton wavelength. For an electron, it has a value of 2.43×10^{-12} m. When the wavelength of the light is long (for instance, the wavelength of the visible light is 0.4–0.7 μm) the wavelength shift becomes negligible. Such interaction between the light and free electrons is called Thomson scattering or Linear Thomson scattering.

The relative strength of the electromagnetic interaction between two charged particles, such as an electron and a proton, is given by the fine-structure constant. This value is a dimensionless quantity formed by the ratio of two energies: the electrostatic energy of attraction (or repulsion) at a separation of one Compton wavelength, and the rest energy of the charge. It is given by $\alpha \approx 7.297353 \times 10^{-3}$, which is approximately equal to 1/137.

When electrons and positrons collide, they annihilate each other, giving rise to two or more gamma ray photons. If the electron and positron have negligible momentum, a positronium atom can form before annihilation results in two or three gamma ray photons totalling 1.022 MeV. On the other hand, high-energy photons may transform into an electron and a positron by a process called pair production, but only in the presence of a nearby charged particle, such as a nucleus.

In the theory of electroweak interaction, the left-handed component of electron's wavefunction forms a weak isospin doublet with the electron neutrino. This means that during weak interactions, electron neutrinos behave like electrons. Either member of this doublet can undergo a charged current interaction by emitting or absorbing a W and be converted into the other member. Charge is conserved during this reaction because the W boson also carries a charge, canceling out any net change during the transmutation. Charged current interactions are responsible for the phenomenon of beta decay in a radioactive atom. Both the electron and electron neutrino can undergo a neutral current interaction via a Z0 exchange, and this is responsible for neutrino-electron elastic scattering.

Atoms and Molecules

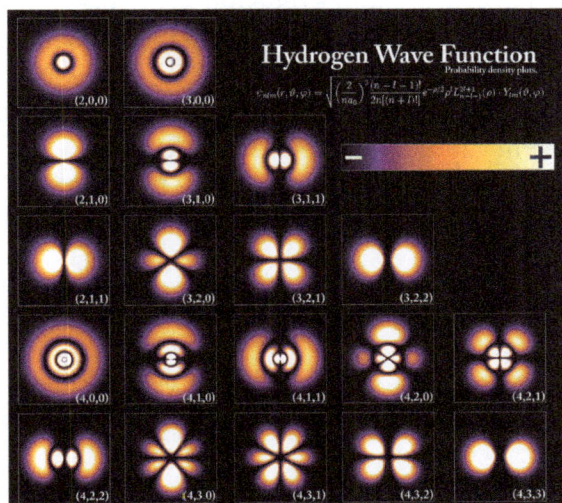

Probability densities for the first few hydrogen atom orbitals, seen in cross-section. The energy level of a bound electron determines the orbital it occupies, and the color reflects the probability of finding the electron at a given position.

An electron can be *bound* to the nucleus of an atom by the attractive Coulomb force. A system of one or more electrons bound to a nucleus is called an atom. If the number of electrons is different from the nucleus' electrical charge, such an atom is called an ion. The wave-like behavior of a bound electron is described by a function called an atomic orbital. Each orbital has its own set of quantum numbers such as energy, angular momentum and projection of angular momentum, and only a discrete set of these orbitals exist around the nucleus. According to the Pauli exclusion principle each orbital can be occupied by up to two electrons, which must differ in their spin quantum number.

Electrons can transfer between different orbitals by the emission or absorption of photons with an energy that matches the difference in potential. Other methods of orbital transfer include collisions with particles, such as electrons, and the Auger effect. To escape the atom, the energy of the electron must be increased above its binding energy to the atom. This occurs, for example, with the photoelectric effect, where an incident photon exceeding the atom's ionization energy is absorbed by the electron.

The orbital angular momentum of electrons is quantized. Because the electron is charged, it produces an orbital magnetic moment that is proportional to the angular momentum. The net magnetic moment of an atom is equal to the vector sum of orbital and spin magnetic moments of all electrons and the nucleus. The magnetic moment of the nucleus is negligible compared with that of the electrons. The magnetic moments of the electrons that occupy the same orbital (so called, paired electrons) cancel each other out.

The chemical bond between atoms occurs as a result of electromagnetic interactions, as described by the laws of quantum mechanics. The strongest bonds are formed by the sharing or transfer of electrons between atoms, allowing the formation of molecules. Within a molecule, electrons move under the influence of several nuclei, and occupy molecular orbitals; much as they can occupy atomic orbitals in isolated atoms. A fundamental factor in these molecular structures is the existence of electron pairs. These are electrons with opposed spins, allowing them to occupy the same molecular orbital without violating the Pauli exclusion principle (much like in atoms). Different molecular orbitals have different spatial distribution of the electron density. For instance, in bonded pairs (i.e. in the pairs that actually bind atoms together) electrons can be found with the maximal probability in a relatively small volume between the nuclei. On the contrary, in non-bonded pairs electrons are distributed in a large volume around nuclei.

Conductivity

A lightning discharge consists primarily of a flow of electrons. The electric potential needed for lightning may be generated by a triboelectric effect.

If a body has more or fewer electrons than are required to balance the positive charge of the nuclei, then that object has a net electric charge. When there is an excess of electrons, the object is said to be negatively charged. When there are fewer electrons than the number of protons in nuclei, the object is said to be positively charged. When the number of electrons and the number of protons are equal, their charges cancel each other and the object is said to be electrically neutral. A macroscopic body can develop an electric charge through rubbing, by the triboelectric effect.

Independent electrons moving in vacuum are termed *free* electrons. Electrons in metals also behave as if they were free. In reality the particles that are commonly termed electrons in metals and other solids are quasi-electrons—quasiparticles, which have the same electrical charge, spin and magnetic moment as real electrons but may have a different mass. When free electrons—both in vacuum and metals—move, they produce a net flow of charge called an electric current, which generates a magnetic field. Likewise a current can be created by a changing magnetic field. These interactions are described mathematically by Maxwell's equations.

At a given temperature, each material has an electrical conductivity that determines the value of electric current when an electric potential is applied. Examples of good conductors include metals such as copper and gold, whereas glass and Teflon are poor conductors. In any dielectric material, the electrons remain bound to their respective atoms and the material behaves as an insulator. Most semiconductors have a variable level of conductivity that lies between the extremes of conduction and insulation. On the other hand, metals have an electronic band structure containing partially filled electronic bands. The presence of such bands allows electrons in metals to behave as if they were free or delocalized electrons. These electrons are not associated with specific atoms, so when an electric field is applied, they are free to move like a gas (called Fermi gas) through the material much like free electrons.

Because of collisions between electrons and atoms, the drift velocity of electrons in a conductor is on the order of millimeters per second. However, the speed at which a change of current at one point in the material causes changes in currents in other parts of the material, the velocity of propagation, is typically about 75% of light speed. This occurs because electrical signals propagate as a wave, with the velocity dependent on the dielectric constant of the material.

Metals make relatively good conductors of heat, primarily because the delocalized electrons are free to transport thermal energy between atoms. However, unlike electrical conductivity, the thermal conductivity of a metal is nearly independent of temperature. This is expressed mathematically by the Wiedemann–Franz law, which states that the ratio of thermal conductivity to the electrical conductivity is proportional to the temperature. The thermal disorder in the metallic lattice increases the electrical resistivity of the material, producing a temperature dependence for electric current.

When cooled below a point called the critical temperature, materials can undergo a phase transition in which they lose all resistivity to electric current, in a process known as superconductivity. In BCS theory, this behavior is modeled by pairs of electrons entering a quantum state known as a Bose–Einstein condensate. These Cooper pairs have their motion coupled to nearby matter via lattice vibrations called phonons, thereby avoiding the collisions with atoms that normally create electrical resistance. (Cooper pairs have a radius of roughly 100 nm, so they can overlap each other.) However, the mechanism by which higher temperature superconductors operate remains uncertain.

Electrons inside conducting solids, which are quasi-particles themselves, when tightly confined at temperatures close to absolute zero, behave as though they had split into three other quasiparticles: spinons, orbitons and holons. The former carries spin and magnetic moment, the next carries its orbital location while the latter electrical charge.

Motion and Energy

According to Einstein's theory of special relativity, as an electron's speed approaches the speed of light, from an observer's point of view its relativistic mass increases, thereby making it more and more difficult to accelerate it from within the observer's frame of reference. The speed of an electron can approach, but never reach, the speed of light in a vacuum, c. However, when relativistic electrons—that is, electrons moving at a speed close to c—are injected into a dielectric medium such as water, where the local speed of light is significantly less than c, the electrons temporarily travel faster than light in the medium. As they interact with the medium, they generate a faint light called Cherenkov radiation.

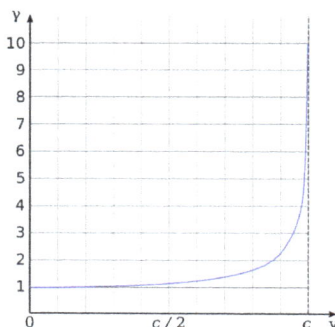

Lorentz factor as a function of velocity. It starts at value 1 and goes to infinity as v approaches c.

The effects of special relativity are based on a quantity known as the Lorentz factor, defined as where v is the speed of the particle. The kinetic energy K_e of an electron moving with velocity v is:

$$K_e = (\gamma - 1)m_e c^2,$$

where m_e is the mass of electron. For example, the Stanford linear accelerator can accelerate an electron to roughly 51 GeV. Since an electron behaves as a wave, at a given velocity it has a characteristic de Broglie wavelength. This is given by $\lambda_e = h/p$ where h is the Planck constant and p is the momentum. For the 51 GeV electron above, the wavelength is about 2.4×10^{-17} m, small enough to explore structures well below the size of an atomic nucleus.

Formation

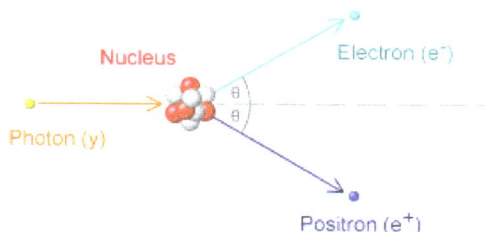

Pair production caused by the collision of a photon with an atomic nucleus

The Big Bang theory is the most widely accepted scientific theory to explain the early stages in the evolution of the Universe. For the first millisecond of the Big Bang, the temperatures were over 10 billion Kelvin and photons had mean energies over a million electronvolts. These photons were sufficiently energetic that they could react with each other to form pairs of electrons and positrons. Likewise, positron-electron pairs annihilated each other and emitted energetic photons:

$$\gamma + \gamma \leftrightarrow e+ + e-$$

An equilibrium between electrons, positrons and photons was maintained during this phase of the evolution of the Universe. After 15 seconds had passed, however, the temperature of the universe dropped below the threshold where electron-positron formation could occur. Most of the surviving electrons and positrons annihilated each other, releasing gamma radiation that briefly reheated the universe.

For reasons that remain uncertain, during the process of leptogenesis there was an excess in the number of electrons over positrons. Hence, about one electron in every billion survived the annihilation process. This excess matched the excess of protons over antiprotons, in a condition known as baryon asymmetry, resulting in a net charge of zero for the universe. The surviving protons and neutrons began to participate in reactions with each other—in the process known as nucleosynthesis, forming isotopes of hydrogen and helium, with trace amounts of lithium. This process peaked after about five minutes. Any leftover neutrons underwent negative beta decay with a half-life of about a thousand seconds, releasing a proton and electron in the process,

$$n \rightarrow p + e- + \nu_e$$

For about the next 300000–400000 years, the excess electrons remained too energetic to bind with atomic nuclei. What followed is a period known as recombination, when neutral atoms were formed and the expanding universe became transparent to radiation.

Roughly one million years after the big bang, the first generation of stars began to form. Within a star, stellar nucleosynthesis results in the production of positrons from the fusion of atomic nuclei. These antimatter particles immediately annihilate with electrons, releasing gamma rays. The net result is a steady reduction in the number of electrons, and a matching increase in the number of neutrons. However, the process of stellar evolution can result in the synthesis of radioactive isotopes. Selected isotopes can subsequently undergo negative beta decay, emitting an electron and antineutrino from the nucleus. An example is the cobalt-60 (^{60}Co) isotope, which decays to form nickel-60 (60Ni).

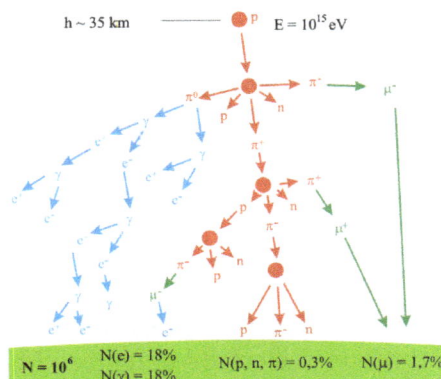

An extended air shower generated by an energetic cosmic ray striking the Earth's atmosphere

At the end of its lifetime, a star with more than about 20 solar masses can undergo gravitational collapse to form a black hole. According to classical physics, these massive stellar objects exert a gravitational attraction that is strong enough to prevent anything, even electromagnetic radiation, from escaping past the Schwarzschild radius. However, quantum mechanical effects are believed to potentially allow the emission of Hawking radiation at this distance. Electrons (and positrons) are thought to be created at the event horizon of these stellar remnants.

When pairs of virtual particles (such as an electron and positron) are created in the vicinity of the event horizon, the random spatial distribution of these particles may permit one of them to appear on the exterior; this process is called quantum tunnelling. The gravitational potential of the black hole can then supply the energy that transforms this virtual particle into a real particle, allowing it to radiate away into space. In exchange, the other member of the pair is given negative energy, which results in a net loss of mass-energy by the black hole. The rate of Hawking radiation increases with decreasing mass, eventually causing the black hole to evaporate away until, finally, it explodes.

Cosmic rays are particles traveling through space with high energies. Energy events as high as 3.0×10^{20} eV have been recorded. When these particles collide with nucleons in the Earth's atmosphere, a shower of particles is generated, including pions. More than half of the cosmic radiation observed from the Earth's surface consists of muons. The particle called a muon is a lepton produced in the upper atmosphere by the decay of a pion.

$$\pi^- \rightarrow \mu^- + \nu\mu$$

A muon, in turn, can decay to form an electron or positron.

$$\mu^- \rightarrow e^- + \nu e + \nu\mu$$

Observation

Aurorae are mostly caused by energetic electrons precipitating into the atmosphere.

Remote observation of electrons requires detection of their radiated energy. For example, in high-energy environments such as the corona of a star, free electrons form a plasma that radiates energy due to Bremsstrahlung radiation. Electron gas can undergo plasma oscillation, which is waves caused by synchronized variations in electron density, and these produce energy emissions that can be detected by using radio telescopes.

The frequency of a photon is proportional to its energy. As a bound electron transitions between different energy levels of an atom, it absorbs or emits photons at characteristic frequencies. For instance, when atoms are irradiated by a source with a broad spectrum, distinct absorption lines appear in the spectrum of transmitted radiation. Each element or molecule displays a characteristic set of spectral lines, such as the hydrogen spectral series. Spectroscopic measurements of the strength and width of these lines allow the composition and physical properties of a substance to be determined.

In laboratory conditions, the interactions of individual electrons can be observed by means of particle detectors, which allow measurement of specific properties such as energy, spin and charge. The development of the Paul trap and Penning trap allows charged particles to be contained within a small region for long durations. This enables precise measurements of the particle properties. For example, in one instance a Penning trap was used to contain a single electron for a period of 10 months. The magnetic moment of the electron was measured to a precision of eleven digits, which, in 1980, was a greater accuracy than for any other physical constant.

The first video images of an electron's energy distribution were captured by a team at Lund University in Sweden, February 2008. The scientists used extremely short flashes of light, called attosecond pulses, which allowed an electron's motion to be observed for the first time.

The distribution of the electrons in solid materials can be visualized by angle-resolved photoemission spectroscopy (ARPES). This technique employs the photoelectric effect to measure the reciprocal space—a mathematical representation of periodic structures that is used to infer the original structure. ARPES can be used to determine the direction, speed and scattering of electrons within the material.

Plasma Applications

Particle Beams

During a NASA wind tunnel test, a model of the Space Shuttle is targeted by a beam of electrons, simulating the effect of ionizing gases during re-entry.

Electron beams are used in welding. They allow energy densities up to 10^7 W·cm^{-2} across a narrow focus diameter of 0.1–1.3 mm and usually require no filler material. This welding technique must be performed in a vacuum to prevent the electrons from interacting with the gas before reaching their target, and it can be used to join conductive materials that would otherwise be considered unsuitable for welding.

Electron-beam lithography (EBL) is a method of etching semiconductors at resolutions smaller than a micrometer. This technique is limited by high costs, slow performance, the need to operate the beam in the vacuum and the tendency of the electrons to scatter in solids. The last problem limits the resolution to about 10 nm. For this reason, EBL is primarily used for the production of small numbers of specialized integrated circuits.

Electron beam processing is used to irradiate materials in order to change their physical properties or sterilize medical and food products. Electron beams fluidise or quasi-melt glasses without significant increase of temperature on intensive irradiation: e.g. intensive electron radiation causes a many orders of magnitude decrease of viscosity and stepwise decrease of its activation energy.

Linear particle accelerators generate electron beams for treatment of superficial tumors in radiation therapy. Electron therapy can treat such skin lesions as basal-cell carcinomas because an electron beam only penetrates to a limited depth before being absorbed, typically up to 5 cm for electron energies in the range 5–20 MeV. An electron beam can be used to supplement the treatment of areas that have been irradiated by X-rays.

Particle accelerators use electric fields to propel electrons and their antiparticles to high energies. These particles emit synchrotron radiation as they pass through magnetic fields. The dependency of the intensity of this radiation upon spin polarizes the electron beam—a process known as the Sokolov–Ternov effect.[note 8] Polarized electron beams can be useful for various experiments. Synchrotron radiation can also cool the electron beams to reduce the momentum spread of the particles. Electron and positron beams are collided upon the particles' accelerating to the required energies; particle detectors observe the resulting energy emissions, which particle physics studies .

Imaging

Low-energy electron diffraction (LEED) is a method of bombarding a crystalline material with a collimated beam of electrons and then observing the resulting diffraction patterns to determine the structure of the material. The required energy of the electrons is typically in the range 20–200 eV. The reflection high-energy electron diffraction (RHEED) technique uses the reflection of a beam of electrons fired at various low angles to characterize the surface of crystalline materials. The beam energy is typically in the range 8–20 keV and the angle of incidence is 1–4°.

The electron microscope directs a focused beam of electrons at a specimen. Some electrons change their properties, such as movement direction, angle, and relative phase and energy as the beam interacts with the material. Microscopists can record these changes in the electron beam to produce atomically resolved images of the material. In blue light, conventional optical microscopes have a diffraction-limited resolution of about 200 nm. By comparison, electron microscopes are limited by the de Broglie wavelength of the electron. This wavelength, for example, is equal to 0.0037 nm for electrons accelerated across a 100,000-volt potential. The Transmission Electron Aberration-Corrected Microscope is capable of sub-0.05 nm resolution, which is more than enough to resolve individual atoms. This capability makes the electron microscope a useful laboratory instrument for high resolution imaging. However, electron microscopes are expensive instruments that are costly to maintain.

Two main types of electron microscopes exist: transmission and scanning. Transmission electron microscopes function like overhead projectors, with a beam of electrons passing through a slice of material then being projected by lenses on a photographic slide or a charge-coupled device. Scanning electron microscopes rasteri a finely focused electron beam, as in a TV set, across the studied sample to produce the image. Magnifications range from 100× to 1,000,000× or higher for both microscope types. The scanning tunneling microscope uses quantum tunneling of electrons from a sharp metal tip into the studied material and can produce atomically resolved images of its surface.

Other Applications

In the free-electron laser (FEL), a relativistic electron beam passes through a pair of undulators that contain arrays of dipole magnets whose fields point in alternating directions. The electrons emit synchrotron radiation that coherently interacts with the same electrons to strongly amplify the radiation field at the resonance frequency. FEL can emit a coherent high-brilliance electromagnetic radiation with a wide range of frequencies, from microwaves to soft X-rays. These devices may find manufacturing, communication and various medical applications, such as soft tissue surgery.

Electrons are important in cathode ray tubes, which have been extensively used as display devices in laboratory instruments, computer monitors and television sets. In a photomultiplier tube, every photon striking the photocathode initiates an avalanche of electrons that produces a detectable current pulse. Vacuum tubes use the flow of electrons to manipulate electrical signals, and they played a critical role in the development of electronics technology. However, they have been largely supplanted by solid-state devices such as the transistor.

Proton

A proton is a subatomic particle, symbol p or p+, with a positive electric charge of +1e elementary charge and mass slightly less than that of a neutron. Protons and neutrons, each with masses of approximately one atomic mass unit, are collectively referred to as "nucleons". One or more protons are present in the nucleus of every atom. The number of protons in the nucleus is the defining property of an element, and is referred to as the atomic number (represented by the symbol Z). Since each element has a unique number of protons, each element has its own unique atomic number. The word *proton* is Greek for "first", and this name was given to the hydrogen nucleus by Ernest Rutherford in 1920. In previous years Rutherford had discovered that the hydrogen nucleus (known to be the lightest nucleus) could be extracted from the nuclei of nitrogen by collision. Protons were therefore a candidate to be a fundamental particle and a building block of nitrogen and all other heavier atomic nuclei.

In the modern Standard Model of particle physics, protons are hadrons, and like neutrons, the other nucleon (particle present in atomic nuclei), are composed of three quarks. Although protons were originally considered fundamental or elementary particles, they are now known to be composed of three valence quarks: two up quarks and one down quark. The rest masses of quarks contribute only about 1% of a proton's mass, however. The remainder of a proton's mass is due to quantum chromodynamics binding energy, which includes the kinetic energy of the quarks and the energy of the gluon fields that bind the quarks together. Because protons are not fundamental particles, they possess a physical size; the radius of a proton is about 0.84–0.87 fm or 0.84×10^{-15} to 0.87×10^{-15} m.

At sufficiently low temperatures, free protons will bind to electrons. However, the character of such bound protons does not change, and they remain protons. A fast proton moving through matter will slow by interactions with electrons and nuclei, until it is captured by the electron cloud of an atom. The result is a protonated atom, which is a chemical compound of hydrogen. In vacuum, when free electrons are present, a sufficiently slow proton may pick up a single free electron, becoming a neutral hydrogen atom, which is chemically a free radical. Such "free hydrogen atoms" tend to react chemically with many other types of atoms at sufficiently low energies. When free hydrogen atoms react with each other, they form neutral hydrogen molecules (H_2), which are the most common molecular component of molecular clouds in interstellar space. Such molecules of hydrogen on Earth may then serve (among many other uses) as a convenient source of protons for accelerators (as used in proton therapy) and other hadron particle physics experiments that require protons to accelerate, with the most powerful and noted example being the Large Hadron Collider.

Description

Protons are spin-½ fermions and are composed of three valence quarks, making them baryons (a sub-type of hadrons). The two up quarks and one down quark of a proton are held together by the strong force, mediated by gluons. A modern perspective has a proton composed of the valence quarks (up, up, down), the gluons, and transitory pairs of sea quarks. Protons have an approximately exponentially decaying positive charge distribution with a mean square radius of about 0.8 fm.

Protons and neutrons are both nucleons, which may be bound together by the nuclear force to form atomic nuclei. The nucleus of the most common isotope of the hydrogen atom (with the chemical symbol "H") is a lone proton. The nuclei of the heavy hydrogen isotopes deuterium and tritium contain one proton bound to one and two neutrons, respectively. All other types of atomic nuclei are composed of two or more protons and various numbers of neutrons.

History

Ernest Rutherford at the first Solvay Conference, 1911

The concept of a hydrogen-like particle as a constituent of other atoms was developed over a long period. As early as 1815, William Prout proposed that all atoms are composed of hydrogen atoms (which he called "protyles"), based on a simplistic interpretation of early values of atomic weights, which was disproved when more accurate values were measured.

In 1886, Eugen Goldstein discovered canal rays (also known as anode rays) and showed that they were positively charged particles (ions) produced from gases. However, since particles from different gases had different values of charge-to-mass ratio (e/m), they could not be identified with a single particle, unlike the negative electrons discovered by J. J. Thomson.

Following the discovery of the atomic nucleus by Ernest Rutherford in 1911, Antonius van den Broek proposed that the place of each element in the periodic table (its atomic number) is equal to its nuclear charge. This was confirmed experimentally by Henry Moseley in 1913 using X-ray spectra.

In 1917 (in experiments reported in 1919), Rutherford proved that the hydrogen nucleus is present in other nuclei, a result usually described as the discovery of protons. Rutherford had earlier learned to produce hydrogen nuclei as a type of radiation produced as a product of the impact of alpha particles on nitrogen gas, and recognize them by their unique penetration signature in air and their appearance in scintillation detectors. These experiments were begun when Rutherford had noticed that, when alpha particles were shot into air (mostly nitrogen), his scintillation detectors showed the signatures of typical hydrogen nuclei as a product. After experimentation Rutherford traced the reaction to the nitrogen in air, and found that when alphas were produced into pure nitrogen gas, the effect was larger. Rutherford determined that this hydrogen could have come only from the nitrogen, and therefore nitrogen must contain hydrogen nuclei. One hydrogen nucleus was being knocked off by the impact of the alpha particle, producing oxygen-17 in the process. This was the first reported nuclear reaction, $^{14}N + \alpha \rightarrow {}^{17}O + p$. (This reaction would later be observed happening directly in a cloud chamber in 1925).

Rutherford knew hydrogen to be the simplest and lightest element and was influenced by Prout's hypothesis that hydrogen was the building block of all elements. Discovery that the hydrogen nucleus is present in all other nuclei as an elementary particle, led Rutherford to give the hydrogen nucleus a special name as a particle, since he suspected that hydrogen, the lightest element, contained only one of these particles. He named this new fundamental building block of the nucleus the *proton*, after the neuter singular of the Greek word for "first", πρ⬚τον. However, Rutherford also had in mind the word *protyle* as used by Prout. Rutherford spoke at the British Association for the Advancement of Science at its Cardiff meeting beginning 24 August 1920. Rutherford was asked by Oliver Lodge for a new name for the positive hydrogen nucleus to avoid confusion with the neutral hydrogen atom. He initially suggested both *proton* and *prouton* (after Prout). Rutherford later reported that the meeting had accepted his suggestion that the hydrogen nucleus be named the "proton", following Prout's word "protyle". The first use of the word "proton" in the scientific literature appeared in 1920.

Recent research has shown that thunderstorms can produce protons with energies of up to several tens of MeV.

Stability

The free proton (a proton not bound to nucleons or electrons) is a stable particle that has not been observed to break down spontaneously to other particles. Free protons are found naturally in a number of situations in which energies or temperatures are high enough to separate them from electrons, for which they have some affinity. Free protons exist in plasmas in which temperatures are too high to allow them to combine with electrons. Free protons of high energy and velocity make up 90% of cosmic rays, which propagate in vacuum for interstellar distances. Free protons are emitted directly from atomic nuclei in some rare types of radioactive decay. Protons also result (along with electrons and antineutrinos) from the radioactive decay of free neutrons, which are unstable.

The spontaneous decay of free protons has never been observed, and protons are therefore considered stable particles according to the Standard Model. However, some grand unified theories (GUTs) of particle physics predict that proton decay should take place with lifetimes between 10^{31} to 10^{36} years and experimental searches have established lower bounds on the mean lifetime of a proton for various assumed decay products.

Experiments at the Super-Kamiokande detector in Japan gave lower limits for proton mean lifetime of 6.6×10^{33} years for decay to an antimuon and a neutral pion, and 8.2×10^{33} years for decay to a positron and a neutral pion. Another experiment at the Sudbury Neutrino Observatory in Canada searched for gamma rays resulting from residual nuclei resulting from the decay of a proton from oxygen-16. This experiment was designed to detect decay to any product, and established a lower limit to a proton lifetime of 2.1×10^{29} years.

However, protons are known to transform into neutrons through the process of electron capture (also called inverse beta decay). For free protons, this process does not occur spontaneously but only when energy is supplied. The equation is:

$$p+ + e- \rightarrow n + ve$$

The process is reversible; neutrons can convert back to protons through beta decay, a common form of radioactive decay. In fact, a free neutron decays this way, with a mean lifetime of about 15 minutes.

Quarks and the Mass of a Proton

In quantum chromodynamics, the modern theory of the nuclear force, most of the mass of protons and neutrons is explained by special relativity. The mass of a proton is about 80–100 times greater than the sum of the rest masses of the quarks that make it up, while the gluons have zero rest mass. The extra energy of the quarks and gluons in a region within a proton, as compared to the rest energy of the quarks alone in the QCD vacuum, accounts for almost 99% of the mass. The rest mass of a proton is, thus, the invariant mass of the system of moving quarks and gluons that make up the particle, and, in such systems, even the energy of massless particles is still measured as part of the rest mass of the system.

Two terms are used in referring to the mass of the quarks that make up protons: *current quark mass* refers to the mass of a quark by itself, while *constituent quark mass* refers to the current quark mass plus the mass of the gluon particle field surrounding the quark. These masses typically have very different values. As noted, most of a proton's mass comes from the gluons that bind the current quarks together, rather than from the quarks themselves. While gluons are inherently massless, they possess energy—to be more specific, quantum chromodynamics binding energy (QCBE)—and it is this that contributes so greatly to the overall mass of protons. A proton has a mass of approximately 938 MeV/c^2, of which the rest mass of its three valence quarks contributes only about 9.4 MeV/c^2; much of the remainder can be attributed to the gluons' QCBE.

The internal dynamics of protons are complicated, because they are determined by the quarks' exchanging gluons, and interacting with various vacuum condensates. Lattice QCD provides a way of calculating the mass of a proton directly from the theory to any accuracy, in principle. The most recent calculations claim that the mass is determined to better than 4% accuracy, even to 1% accuracy. These claims are still controversial, because the calculations cannot yet be done with quarks as light as they are in the real world. This means that the predictions are found by a process of extrapolation, which can introduce systematic errors. It is hard to tell whether these errors are controlled properly, because the quantities that are compared to experiment are the masses of the hadrons, which are known in advance.

These recent calculations are performed by massive supercomputers, and, as noted by Boffi and Pasquini: "a detailed description of the nucleon structure is still missing because … long-distance behavior requires a nonperturbative and/or numerical treatment…" More conceptual approaches to the structure of protons are: the topological soliton approach originally due to Tony Skyrme and the more accurate AdS/QCD approach that extends it to include a string theory of gluons, various QCD-inspired models like the bag model and the constituent quark model, which were popular in the 1980s, and the SVZ sum rules, which allow for rough approximate mass calculations. These methods do not have the same accuracy as the more brute-force lattice QCD methods, at least not yet.

Charge Radius

The problem of defining a radius for an atomic nucleus (proton) is similar to the problem of atomic radius, in that neither atoms nor their nuclei have definite boundaries. However, the nucleus can be modeled as a sphere of positive charge for the interpretation of electron scattering experiments: because there is no definite boundary to the nucleus, the electrons "see" a range of cross-sections, for which a mean can be taken. The qualification of "rms" (for "root mean square") arises because it is the nuclear cross-section, proportional to the square of the radius, which is determining for electron scattering.

The internationally accepted value of a proton's charge radius is 0.8768 fm. This value is based on measurements involving a proton and an electron (namely, electron scattering measurements and complex calculation involving scattering cross section based on Rosenbluth equation for momentum-transfer cross section), and studies of the atomic energy levels of hydrogen and deuterium.

However, in 2010 an international research team published a proton charge radius measurement via the Lamb shift in muonic hydrogen (an exotic atom made of a proton and a negatively charged muon). Their measurement of the root-mean-square charge radius of a proton is "0.84184(67) fm, which differs by 5.0 standard deviations from the CODATA value of 0.8768(69) fm". In January 2013, an updated value for the charge radius of a proton—0.84087(39) fm—was published. The precision was improved by 1.7 times, but the difference with CODATA value persisted at 7σ significance.

The international research team that obtained this result at the Paul Scherrer Institut (PSI) in Villigen (Switzerland) includes scientists from the Max Planck Institute of Quantum Optics (MPQ) in Garching, the Ludwig-Maximilians-Universität (LMU) Munich and the Institut für Strahlwerkzeuge (IFWS) of the Universität Stuttgart (both from Germany), and the University of Coimbra, Portugal.

They are now attempting to explain the discrepancy, and re-examining the results of both previous high-precision measurements and complex calculations involving scattering cross section. If no errors are found in the measurements or calculations, it could be necessary to re-examine the world's most precise and best-tested fundamental theory: quantum electrodynamics. The proton radius remains a puzzle as of early 2015. Perhaps the discrepancy is due to new physics or the explanation is an ordinary physics effect that has been missed.

The radius is linked to the form factor and momentum transfer cross section. The atomic form factor G modifies the cross section corresponding to point-like proton.

$$R_e^2 = -6 \frac{dG_e}{dq^2}\bigg|_{q^2=0} \frac{d\sigma}{d\Omega} = \frac{d\sigma}{d\Omega}\bigg|_{point} G^2(q^2)$$

The atomic form factor is related to the wave function density of the target:

$$G(q^2) = \int e^{iqr}\psi(r)^2 dr^3$$

The form factor can be split in electric and magnetic form factors. These can be further written as linear combinations of Dirac and Pauli form factors.

$$G_m = F_D + F_P G_e = F_D - \tau F_P \frac{d\sigma}{d\Omega} = \frac{d\sigma}{d\Omega}\bigg|_{NS} \frac{1}{1+\tau}\left(G_e^2(q^2) + \frac{\tau}{\epsilon}G_m^2(q^2)\right)$$

Interaction of Free Protons with Ordinary Matter

Although protons have affinity for oppositely charged electrons, free protons must lose sufficient velocity (and kinetic energy) in order to become closely associated and bound to electrons, since this is a relatively low-energy interaction. High energy protons, in traversing ordinary matter, lose energy by collisions with atomic nuclei, and by ionization of atoms (removing electrons) until they are slowed sufficiently to be captured by the electron cloud in a normal atom.

However, in such an association with an electron, the character of the bound proton is not changed, and it remains a proton. The attraction of low-energy free protons to any electrons present in normal matter (such as the electrons in normal atoms) causes free protons to stop and to form a new chemical bond with an atom. Such a bond happens at any sufficiently "cold" temperature (i.e., comparable to temperatures at the surface of the Sun) and with any type of atom. Thus, in interaction with any type of normal (non-plasma) matter, low-velocity free protons are attracted to electrons in any atom or molecule with which they come in contact, causing the proton and molecule to combine. Such molecules are then said to be "protonated", and chemically they often, as a result, become so-called Bronsted acids.

Proton in Chemistry

Atomic Number

In chemistry, the number of protons in the nucleus of an atom is known as the atomic number, which determines the chemical element to which the atom belongs. For example, the atomic number of chlorine is 17; this means that each chlorine atom has 17 protons and that all atoms with 17 protons are chlorine atoms. The chemical properties of each atom are determined by the number of (negatively charged) electrons, which for neutral atoms is equal to the number of (positive) protons so that the total charge is zero. For example, a neutral chlorine atom has 17 protons and 17 electrons, whereas a Cl^- anion has 17 protons and 18 electrons for a total charge of −1.

All atoms of a given element are not necessarily identical, however, as the number of neutrons may vary to form different isotopes, and energy levels may differ forming different nuclear isomers. For example, there are two stable isotopes of chlorine: 35 17Cl with 35 − 17 = 18 neutrons and 37 17Cl with 37 − 17 = 20 neutrons.

Hydrogen Ion

Protium, the most common isotope of hydrogen, consists of one proton and one electron (it has no neutrons). The term "hydrogen ion" (H+) implies that that H-atom has lost its one electron, causing only a proton to remain. Thus, in chemistry, the terms "proton" and "hydrogen ion" (for the protium isotope) are used synonymously

In chemistry, the term proton refers to the hydrogen ion, H+. Since the atomic number of hydrogen is 1, a hydrogen ion has no electrons and corresponds to a bare nucleus, consisting of a proton (and 0 neutrons for the most abundant isotope *protium* 11H). The proton is a "bare charge" with only about 1/64,000 of the radius of a hydrogen atom, and so is extremely reactive chemically. The free proton, thus, has an extremely short lifetime in chemical systems such as liquids and it reacts immediately with the electron cloud of any available molecule. In aqueous solution, it forms the hydronium ion, H_3O^+, which in turn is further solvated by water molecules in clusters such as $[H_5O_2]^+$ and $[H_9O_4]^+$.

The transfer of H+ in an acid–base reaction is usually referred to as "proton transfer". The acid is referred to as a proton donor and the base as a proton acceptor. Likewise, biochemical terms such as proton pump and proton channel refer to the movement of hydrated H+ ions.

The ion produced by removing the electron from a deuterium atom is known as a deuteron, not a proton. Likewise, removing an electron from a tritium atom produces a triton.

Proton Nuclear Magnetic Resonance (NMR)

Also in chemistry, the term "proton NMR" refers to the observation of hydrogen-1 nuclei in (mostly organic) molecules by nuclear magnetic resonance. This method uses the spin of the proton, which has the value one-half. The name refers to examination of protons as they occur in protium (hydrogen-1 atoms) in compounds, and does not imply that free protons exist in the compound being studied.

Human Exposure

The Apollo Lunar Surface Experiments Packages (ALSEP) determined that more than 95% of the particles in the solar wind are electrons and protons, in approximately equal numbers.

Because the Solar Wind Spectrometer made continuous measurements, it was possible to measure how the Earth's magnetic field affects arriving solar wind particles. For about two-thirds of each orbit, the Moon is outside of the Earth's magnetic field. At these times, a typical proton density was 10 to 20 per cubic centimeter, with most protons having velocities between 400 and 650 kilometers per second. For about five days of each month, the Moon is inside the Earth's geomagnetic tail, and typically no solar wind particles were detectable. For the remainder of each lunar orbit, the Moon is in a transitional region known as the magnetosheath, where the Earth's magnetic field affects the solar wind but does not completely exclude it. In this region, the particle flux is reduced, with typical proton velocities of 250 to 450 kilometers per second. During the lunar night, the spectrometer was shielded from the solar wind by the Moon and no solar wind particles were measured.

Protons also have extrasolar origin from galactic cosmic rays, where they make up about 90% of the total particle flux. These protons often have higher energy than solar wind protons, and their intensity is far more uniform and less variable than protons coming from the Sun, the production of which is heavily affected by solar proton events such as coronal mass ejections.

Research has been performed on the dose-rate effects of protons, as typically found in space travel, on human health. To be more specific, there are hopes to identify what specific chromosomes are damaged, and to define the damage, during cancer development from proton exposure. Another study looks into determining "the effects of exposure to proton irradiation on neurochemical and behavioral endpoints, including dopaminergic functioning, amphetamine-induced conditioned taste aversion learning, and spatial learning and memory as measured by the Morris water maze. Electrical charging of a spacecraft due to interplanetary proton bombardment has also been proposed for study. There are many more studies that pertain to space travel, including galactic cosmic rays and their possible health effects, and solar proton event exposure.

The American Biostack and Soviet Biorack space travel experiments have demonstrated the severity of molecular damage induced by heavy ions on micro organisms including Artemia cysts.

Antiproton

CPT-symmetry puts strong constraints on the relative properties of particles and antiparticles and, therefore, is open to stringent tests. For example, the charges of a proton and antiproton must sum to exactly zero. This equality has been tested to one part in 10^8. The equality of their masses has also been tested to better than one part in 10^8. By holding antiprotons in a Penning trap, the equality of the charge to mass ratio of protons and antiprotons has been tested to one part in 6×10^9. The magnetic moment of antiprotons has been measured with error of 8×10^{-3} nuclear Bohr magnetons, and is found to be equal and opposite to that of a proton.

Neutron

The neutron is a subatomic particle, symbol n or n0, with no net electric charge and a mass slightly larger than that of a proton. Protons and neutrons, each with mass approximately one atomic mass unit, constitute the nucleus of an atom, and they are collectively referred to as nucleons. Their properties and interactions are described by nuclear physics.

The nucleus consists of Z protons, where Z is called the atomic number, and N neutrons, where N is the neutron number. The atomic number defines the chemical properties of the atom, and the neutron number determines the isotope or nuclide. The terms isotope and nuclide are often used synonymously, but they are chemical and nuclear concepts, respectively. The atomic mass number, symbol A, equals Z+N. For example, carbon has atomic number 6, and its abundant carbon-12 isotope has 6 neutrons, whereas its rare carbon-13 isotope has 7 neutrons. Some elements occur in nature with only one stable isotope, such as fluorine. Other elements occur as many stable isotopes, such as tin with ten stable isotopes. Even though it is not a chemical element, the neutron is included in the table of nuclides.

Within the nucleus, protons and neutrons are bound together through the nuclear force, and neutrons are required for the stability of nuclei. Neutrons are produced copiously in nuclear fission and fusion. They are a primary contributor to the nucleosynthesis of chemical elements within stars through fission, fusion, and neutron capture processes.

The neutron is essential to the production of nuclear power. In the decade after the neutron was discovered in 1932, neutrons were used to induce many different types of nuclear transmutations. With the discovery of nuclear fission in 1938, it was quickly realized that, if a fission event produced neutrons, each of these neutrons might cause further fission events, etc., in a cascade known as a nuclear chain reaction. These events and findings led to the first self-sustaining nuclear reactor (Chicago Pile-1, 1942) and the first nuclear weapon (Trinity, 1945).

Free neutrons, or individual neutrons free of the nucleus, are effectively a form of ionizing radiation, and as such, are a biological hazard, depending upon dose. A small natural "neutron background" flux of free neutrons exists on Earth, caused by cosmic ray showers, and by the natural radioactivity of spontaneously fissionable elements in the Earth's crust. Dedicated neutron sources like neutron generators, research reactors and spallation sources produce free neutrons for use in irradiation and in neutron scattering experiments.

Description

Neutrons and protons are both nucleons, which are attracted and bound together by the nuclear force to form atomic nuclei. The nucleus of the most common isotope of the hydrogen atom (with the chemical symbol "H") is a lone proton. The nuclei of the heavy hydrogen isotopes deuterium and tritium contain one proton bound to one and two neutrons, respectively. All other types of atomic nuclei are composed of two or more protons and various numbers of neutrons. The most common nuclide of the common chemical element lead, ^{208}Pb has 82 protons and 126 neutrons, for example.

The free neutron has a mass of about 1.675×10^{-27} kg (equivalent to 939.6 MeV/c^2, or 1.0087 u). The neutron has a mean square radius of about 0.8×10^{-15} m, or 0.8 fm, and it is a spin-½ fermion. The neutron has a magnetic moment with a negative value, because its orientation is opposite to the neutron's spin. The neutron's magnetic moment causes its motion to be influenced by magnetic fields. Although the neutron has no net electric charge, it does have a slight distribution of charge within it. With its positive electric charge, the proton is directly influenced by electric fields, whereas the response of the neutron to this force is much weaker.

A free neutron is unstable, decaying to a proton, electron and antineutrino with a mean lifetime of just under 15 minutes (881.5 ± 1.5 s). This radioactive decay, known as beta decay, is possible since the mass of the neutron is slightly greater than the proton. The free proton is stable. Neutrons or protons bound in a nucleus can be stable or unstable, however, depending on the nuclide. Beta decay, in which neutrons decay to protons, or vice versa, is governed by the weak force, and it requires the emission or absorption of electrons and neutrinos, or their antiparticles.

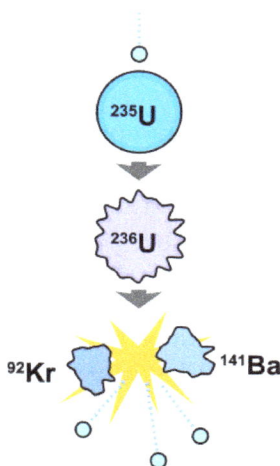

Nuclear fission caused by absorption of a neutron by uranium-235. The heavy nuclide fragments into lighter components and additional neutrons.

Protons and neutrons behave almost identically under the influence of the nuclear force within the nucleus. The concept of isospin, in which the proton and neutron are viewed as two quantum states of the same particle, is used to model the interactions of nucleons by the nuclear or weak forces. Because of the strength of the nuclear force at short distances, the binding energy of nucleons is more than seven orders of magnitude larger than the electromagnetic energy binding electrons in atoms. Nuclear reactions (such as nuclear fission) therefore have an energy density that is more than ten million times that of chemical reactions. Because of the mass–energy equivalence, nuclear binding energies add or subtract from the mass of nuclei. Ultimately, the ability of the nuclear force to store energy arising from the electromagnetic repulsion of nuclear components is the basis for most of the energy that makes nuclear reactors or bombs possible. In nuclear fission, the absorption of a neutron by a heavy nuclide (e.g., uranium-235) causes the nuclide to become unstable and break into light nuclides and additional neutrons. The positively charged light nuclides then repel, releasing electromagnetic potential energy.

The neutron is classified as a hadron, since it is composed of quarks, and as a baryon, since it is composed of three quarks. The finite size of the neutron and its magnetic moment indicate the neutron is a composite, rather than elementary, particle. The neutron consists of two down quarks with charge $-\frac{1}{3}$ e and one up quark with charge $+\frac{2}{3}$ e, although this simple model belies the complexities of the Standard Model for nuclei. The masses of the three quarks sum to only about 12 MeV/c^2, whereas the neutron's mass is about 940 MeV/c^2, for example. Like the proton, the quarks of the neutron are held together by the strong force, mediated by gluons. The nuclear force results from secondary effects of the more fundamental strong force.

Discovery

The story of the discovery of the neutron and its properties is central to the extraordinary developments in atomic physics that occurred in the first half of the 20th century, leading ultimately to the atomic bomb in 1945. In the 1911 Rutherford model, the atom consisted of a small positively charged massive nucleus surrounded by a much larger cloud of negatively charged electrons. In 1920, Rutherford suggested that the nucleus consisted of positive protons and neutrally-charged particles, suggested to be a proton and an electron bound in some way. Electrons were assumed to reside within the nucleus because it was known that beta radiation consisted of electrons emitted from the nucleus. Rutherford called these uncharged particles *neutrons*, by the Latin root for *neutralis* (neuter) and the Greek suffix -*on* (a suffix used in the names of subatomic particles, i.e. *electron* and *proton*). References to the word *neutron* in connection with the atom can be found in the literature as early as 1899, however.

Throughout the 1920s, physicists assumed that the atomic nucleus was composed of protons and "nuclear electrons" but there were obvious problems. It was difficult to reconcile the proton–electron model for nuclei with the Heisenberg uncertainty relation of quantum mechanics. The Klein paradox, discovered by Oskar Klein in 1928, presented further quantum mechanical objections to the notion of an electron confined within a nucleus. Observed properties of atoms and molecules were inconsistent with the nuclear spin expected from proton–electron hypothesis. Since both protons and electrons carry an intrinsic spin of ½ $\hbar$, there is no way to arrange an odd number of spins ±½ $\hbar$ to give a spin integer multiple of $\hbar$. Nuclei with integer spin are common, e.g., ^{14}N.

In 1931, Walther Bothe and Herbert Becker found that if alpha particle radiation from polonium fell on beryllium, boron, or lithium, an unusually penetrating radiation was produced. The radiation was not influenced by an electric field, so Bothe and Becker assumed it was gamma radiation. The following year Irène Joliot-Curie and Frédéric Joliot in Paris showed that if this "gamma" radiation fell on paraffin, or any other hydrogen-containing compound, it ejected protons of very high energy. Neither Rutherford nor James Chadwick at the Cavendish Laboratory in Cambridge were convinced by the gamma ray interpretation. Chadwick quickly performed a series of experiments that showed that the new radiation consisted of uncharged particles with about the same mass as the proton. These particles were neutrons. Chadwick won the Nobel Prize in Physics for this discovery in 1935.

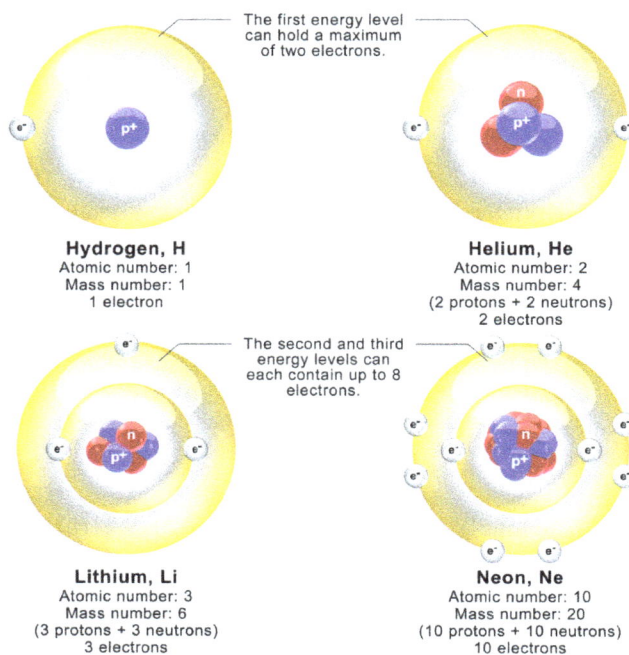

Models depicting the nucleus and electron energy levels in hydrogen, helium, lithium, and neon atoms. In reality, the diameter of the nucleus is about 100,000 times smaller than the diameter of the atom.

Models for atomic nucleus consisting of protons and neutrons were quickly developed by Werner Heisenberg and others. The proton–neutron model explained the puzzle of nuclear spins. The origins of beta radiation were explained by Enrico Fermi in 1934 by the process of beta decay, in which the neutron decays to a proton by *creating* an electron and a (as yet undiscovered) neutrino. In 1935 Chadwick and his doctoral student Maurice Goldhaber, reported the first accurate measurement of the mass of the neutron.

By 1934, Fermi had bombarded heavier elements with neutrons to induce radioactivity in elements of high atomic number. In 1938, Fermi received the Nobel Prize in Physics *"for his demonstrations of the existence of new radioactive elements produced by neutron irradiation, and for his related discovery of nuclear reactions brought about by slow neutrons"*. In 1938 Otto Hahn, Lise Meitner, and Fritz Strassmann discovered nuclear fission, or the fractionation of uranium nuclei into light elements, induced by neutron bombardment. In 1945 Hahn received the 1944 Nobel Prize in Chemistry *"for his discovery of the fission of heavy atomic nuclei."* The discovery of nuclear fission would lead to the development of nuclear power and the atomic bomb by the end of World War II.

Beta Decay and the Stability of the Nucleus

Under the Standard Model of particle physics, the only possible decay mode for the neutron that conserves baryon number is for one of the neutron's quarks to change flavour via the weak interaction. The decay of one of the neutron's down quarks into a lighter up quark can be achieved by the emission of a W boson. By this process, the Standard Model description of beta decay, the neutron decays into a proton (which contains one down and two up quarks), an electron, and an electron antineutrino.

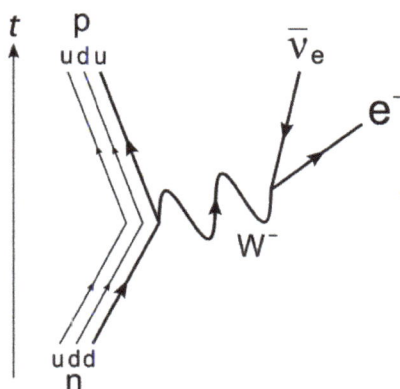

The Feynman diagram for beta decay of a neutron into a proton, electron, and electron antineutrino via an intermediate heavy W boson

Since interacting protons have a mutual electromagnetic repulsion that is stronger than their attractive nuclear interaction, neutrons are a necessary constituent of any atomic nucleus that contains more than one proton. Neutrons bind with protons and one another in the nucleus via the nuclear force, effectively moderating the repulsive forces between the protons and stabilizing the nucleus.

Valley of stability, Beta-decay stable isobars, and Neutron emission

Free neutron decay

Outside the nucleus, free neutrons are unstable and have a mean lifetime of 881.5±1.5 s (about 14 minutes, 42 seconds); therefore the half-life for this process (which differs from the mean lifetime by a factor of ln(2) = 0.693) is 611.0±1.0 s (about 10 minutes, 11 seconds). Beta decay of the neutron, described above, can be denoted by the radioactive decay:

no → p+ + e− + ve

where p+, e−, and ve denote the proton, electron and electron antineutrino, respectively. For the free neutron the decay energy for this process (based on the masses of the neutron, proton, and electron) is 0.782343 MeV. The maximal energy of the beta decay electron (in the process wherein the neutrino receives a vanishingly small amount of kinetic energy) has been measured at 0.782 ± .013 MeV. The latter number is not well-enough measured to determine the comparatively tiny rest mass of the neutrino (which must in theory be subtracted from the maximal electron kinetic energy) as well as neutrino mass is constrained by many other methods.

A small fraction (about one in 1000) of free neutrons decay with the same products, but add an extra particle in the form of an emitted gamma ray:

no → p+ + e− + ve + γ

This gamma ray may be thought of as a sort of "internal bremsstrahlung" that arises as the emitted beta particle interacts with the charge of the proton in an electromagnetic way. Internal bremsstrahlung gamma ray production is also a minor feature of beta decays of bound neutrons (as discussed below).

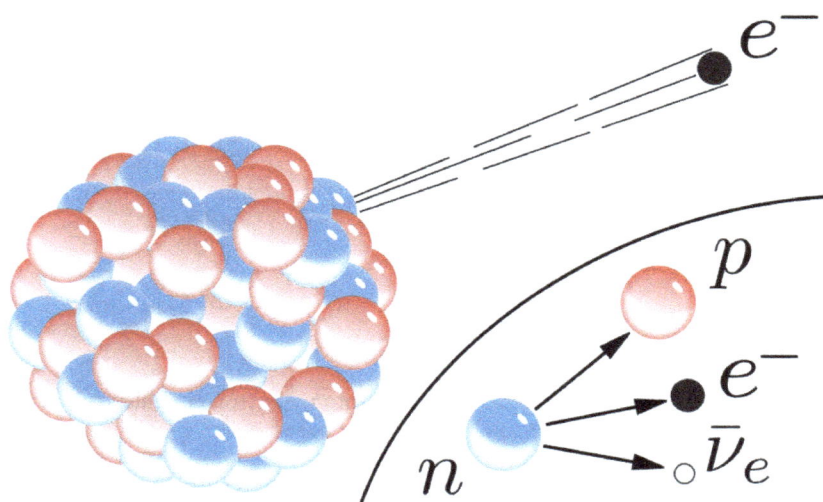

A schematic of the nucleus of an atom indicating β− radiation, the emission of a fast electron from the nucleus (the accompanying antineutrino is omitted). In the Rutherford model for the nucleus, red spheres were protons with positive charge and blue spheres were protons tightly bound to an electron with no net charge.
The **inset** shows beta decay of a free neutron as it is understood today; an electron and antineutrino are created in this process.

A very small minority of neutron decays (about four per million) are so-called "two-body (neutron) decays", in which a proton, electron and antineutrino are produced as usual, but the electron fails to gain the 13.6 eV necessary energy to escape the proton (the ionization energy of hydrogen), and therefore simply remains bound to it, as a neutral hydrogen atom (one of the "two bodies"). In this type of free neutron decay, in essence all of the neutron decay energy is carried off by the antineutrino (the other "body").

The transformation of a free proton to a neutron (plus a positron and a neutrino) is energetically impossible, since a free neutron has a greater mass than a free proton.

Bound Neutron Decay

While a free neutron has a half life of about 10.2 min, most neutrons within nuclei are stable. According to the nuclear shell model, the protons and neutrons of a nuclide are a quantum mechanical system organized into discrete energy levels with unique quantum numbers. For a neutron to decay, the resulting proton requires an available state at lower energy than the initial neutron state. In stable nuclei the possible lower energy states are all filled, meaning they are each occupied by two protons with spin up and spin down. The Pauli exclusion principle therefore disallows the decay of a neutron to a proton within stable nuclei. The situation is similar to electrons of an atom, where electrons have distinct atomic orbitals and are prevented from decaying to lower energy states, with the emission of a photon, by the exclusion principle.

Neutrons in unstable nuclei can decay by beta decay as described above. In this case, an energetically allowed quantum state is available for the proton resulting from the decay. One example of this decay is carbon-14 (6 protons, 8 neutrons) that decays to nitrogen-14 (7 protons, 7 neutrons) with a half-life of about 5,730 years.

Inside a nucleus, a proton can transform into a neutron via inverse beta decay, if an energetically allowed quantum state is available for the neutron. This transformation occurs by emission of an antielectron (also called positron) and an electron neutrino:

p+ → n0 + e+ + ve

The transformation of a proton to a neutron inside of a nucleus is also possible through electron capture:

p+ + e− → n0 + ve

Positron capture by neutrons in nuclei that contain an excess of neutrons is also possible, but is hindered because positrons are repelled by the positive nucleus, and quickly annihilate when they encounter electrons.

Competition of Beta Decay Types

Three types of beta decay in competition are illustrated by the single isotope copper-64 (29 protons, 35 neutrons), which has a half-life of about 12.7 hours. This isotope has one unpaired proton and one unpaired neutron, so either the proton or the neutron can decay. This particular nuclide is almost equally likely to undergo proton decay (by positron emission, 18% or by electron capture, 43%) or neutron decay (by electron emission, 39%).

Intrinsic Properties

Mass

The mass of a neutron cannot be directly determined by mass spectrometry due to lack of electric charge. However, since the mass of protons and deuterons can be measured by mass spectrometry, the mass of a neutron can be deduced by subtracting proton mass from deuteron mass, with the difference being the mass of the neutron plus the binding energy of deuterium (expressed as a positive emitted energy). The latter can be directly measured by measuring the energy (B_d) of the single 0.7822 MeV gamma photon emitted when neutrons are captured by protons (this is exothermic and happens with zero-energy neutrons), plus the small recoil kinetic energy (E_{rd}) of the deuteron (about 0.06% of the total energy).

$$m_n = m_d - m_p + B_d - E_{rd}$$

The energy of the gamma ray can be measured to high precision by X-ray diffraction techniques, as was first done by Bell and Elliot in 1948. The best modern (1986) values for neutron mass by this technique are provided by Greene, et al. These give a neutron mass of:

$$m_{neutron} = 1.008644904(14) \text{ u}$$

The value for the neutron mass in MeV is less accurately known, due to less accuracy in the known conversion of u to MeV:

$$m_{neutron} = 939.56563(28) \text{ MeV}/c^2.$$

Another method to determine the mass of a neutron starts from the beta decay of the neutron, when the momenta of the resulting proton and electron are measured.

Electric Charge

The total electric charge of the neutron is 0 e. This zero value has been tested experimentally, and the present experimental limit for the charge of the neutron is $-2(8) \times 10^{-22}$ e, or $-3(13) \times 10^{-41}$ C. This value is consistent with zero, given the experimental uncertainties (indicated in parentheses). By comparison, the charge of the proton is, of course, $+1$ e.

Magnetic Moment

Even though the neutron is a neutral particle, the magnetic moment of a neutron is not zero. Since the neutron is a neutral particle, it is not affected by electric fields, but with its magnetic moment it is affected by magnetic fields. The magnetic moment of the neutron is an indication of its quark substructure and internal charge distribution. The value for the neutron's magnetic moment was first directly measured by Luis Alvarez and Felix Bloch at Berkeley, California in 1940, using an extension of the magnetic resonance methods developed by Rabi. Alvarez and Bloch determined the magnetic moment of the neutron to be $\mu_n = -1.93(2)$ μ_N, where μ_N is the nuclear magneton.

In the quark model for hadrons, the neutron is composed of one up quark (charge $+1/3$ e) and two down quarks (charge $-1/3$ e). The magnetic moment of the neutron can be modeled as a sum of the magnetic moments of the constituent quarks, although this simple model belies the complexities of the Standard Model of particle physics. The calculation assumes that the quarks behave like pointlike Dirac particles, each having their own magnetic moment. Simplistically, the magnetic moment of the neutron can be viewed as resulting from the vector sum of the three quark magnetic moments, plus the orbital magnetic moments caused by the movement of the three charged quarks within the neutron.

In one of the early successes of the Standard Model (SU(6) theory, now understood in terms of quark behavior), in 1964 Mirza A. B. Beg, Benjamin W. Lee, and Abraham Pais theoretically calculated the ratio of proton to neutron magnetic moments to be $-3/2$, which agrees with the experimental value to within 3%. The measured value for this ratio is $-1.45989806(34)$. A contradiction of the quantum mechanical basis of this calculation with the Pauli exclusion principle, led to the discovery of the color charge for quarks by Oscar W. Greenberg in 1964.

The above treatment compares neutrons with protons, allowing the complex behavior of quarks to be subtracted out between models, and merely exploring what the effects would be of differing quark charges (or quark type). Such calculations are enough to show that the interior of neutrons is very much like that of protons, save for the difference in quark composition with a down quark in the neutron replacing an up quark in the proton.

Attempts have been made to quantitatively recover the neutron magnetic moment from first principles. From the nonrelativistic, quantum mechanical wavefunction for baryons composed of three quarks, a straightforward calculation gives fairly accurate estimates for the magnetic moments of neutrons, protons, and other baryons. For a neutron, the end result of this calculation is that the magnetic moment of the neutron is given by $\mu_n = 4/3\,\mu_d - 1/3\,\mu_u$, where μ_d and μ_u are the magnetic moments for the down and up quarks, respectively. This result combines the intrinsic magnetic moments of the quarks with their orbital magnetic moments, and assumes the three quarks are in a particular, dominant quantum state.

Baryon	Magnetic moment of quark model	Computed (μ_N)	Observed (μ_N)
p	$4/3\,\mu_u - 1/3\,\mu_d$	2.79	2.793
n	$4/3\,\mu_d - 1/3\,\mu_u$	−1.86	−1.913

The results of this calculation are encouraging, but the masses of the up or down quarks were assumed to be 1/3 the mass of a nucleon. The masses of the quarks are actually only about 1% that of a nucleon. The discrepancy stems from the complexity of the Standard Model for nucleons, where most of their mass originates in the gluon fields, virtual particles, and their associated energy that are essential aspects of the strong force. Furthermore, the complex system of quarks and gluons that constitute a neutron requires a relativistic treatment. The nucleon magnetic moment has been successfully computed numerically from first principles, however, including all the effects mentioned and using more realistic values for the quark masses. The calculation gave results that were in fair agreement with measurement, but it required significant computing resources.

Spin

The neutron is a spin 1/2 particle, that is, it is a fermion with intrinsic angular momentum equal to $1/2\,\hbar$, where $\hbar$ is the reduced Planck constant. For many years after the discovery of the neutron, its exact spin was ambiguous. Although it was assumed to be a spin 1/2 Dirac particle, the possibility that the neutron was a spin 3/2 particle had not been ruled out. The interactions of the neutron's magnetic moment with an external magnetic field were exploited to finally determine the spin of the neutron. In 1949, Hughes and Burgy measured neutrons reflected from a ferromagnetic mirror and found that the angular distribution of the reflections was consistent with spin 1/2. In 1954, Sherwood, Stephenson, and Bernstein employed neutrons in a Stern–Gerlach experiment that used a magnetic field to separate the neutron spin states. They recorded two such spin states, consistent with a spin 1/2 particle.

Structure and Geometry of Charge Distribution

An article published in 2007 featuring a model-independent analysis concluded that the neutron has a negatively charged exterior, a positively charged middle, and a negative core. In a simplified classical view, the negative "skin" of the neutron assists it to be attracted to the protons with which it interacts in the nucleus. (However, the main attraction between neutrons and protons is via the nuclear force, which does not involve charge.)

The simplified classical view of the neutron's charge distribution also "explains" the fact that the neutron magnetic dipole points in the opposite direction from its spin angular momentum vector (as compared to the proton). This gives the neutron, in effect, a magnetic moment which resembles a negatively charged particle. This can be reconciled classically with a neutral neutron composed of a charge distribution in which the negative sub-parts of the neutron have a larger average radius of distribution, and therefore contribute more to the particle's magnetic dipole moment, than do the positive parts that are, on average, nearer the core.

Electric Dipole Moment

The Standard Model of particle physics predicts a tiny separation of positive and negative charge within the neutron leading to a permanent electric dipole moment. The predicted value is, however, well below the current sensitivity of experiments. From several unsolved puzzles in particle physics, it is clear that the Standard Model is not the final and full description of all particles and their interactions. New theories going beyond the Standard Model generally lead to much larger predictions for the electric dipole moment of the neutron. Currently, there are at least four experiments trying to measure for the first time a finite neutron electric dipole moment, including:

- Cryogenic neutron EDM experiment being set up at the Institut Laue–Langevin

- nEDM experiment under construction at the new UCN source at the Paul Scherrer Institute

- nEDM experiment being envisaged at the Spallation Neutron Source

- nEDM experiment being built at the Institut Laue–Langevin

Anti-Neutron

The antineutron is the antiparticle of the neutron. It was discovered by Bruce Cork in the year 1956, a year after the antiproton was discovered. CPT-symmetry puts strong constraints on the relative properties of particles and antiparticles, so studying antineutrons yields provide stringent tests on CPT-symmetry. The fractional difference in the masses of the neutron and antineutron is $(9\pm6)\times10^{-5}$. Since the difference is only about two standard deviations away from zero, this does not give any convincing evidence of CPT-violation.

Neutron Compounds

Dineutrons and Tetraneutrons

The existence of stable clusters of 4 neutrons, or tetraneutrons, has been hypothesised by a team led by Francisco-Miguel Marqués at the CNRS Laboratory for Nuclear Physics based on observations of the disintegration of beryllium-14 nuclei. This is particularly interesting because current theory suggests that these clusters should not be stable.

In February 2016, Japanese physicist Susumu Shimoura of the University of Tokyo and co-workers reported they had observed the purported tetraneutrons for the first time experimentally. Nuclear physicists around the world say this discovery, if confirmed, would be a milestone in the field of nuclear physics and certainly would deepen our understanding of the nuclear forces.

The dineutron is another hypothetical particle. In 2012, Artemis Spyrou from Michigan State University and coworkers reported that they observed, for the first time, the dineutron emission in the decay of ^{16}Be. The dineutron character is evidenced by a small emission angle between the two neutrons. The authors measured the two-neutron separation energy to be 1.35(10) MeV, in good agreement with shell model calculations, using standard interactions for this mass region.

Neutronium and Neutron Stars

At extremely high pressures and temperatures, nucleons and electrons are believed to collapse into bulk neutronic matter, called neutronium. This is presumed to happen in neutron stars.

The extreme pressure inside a neutron star may deform the neutrons into a cubic symmetry, allowing tighter packing of neutrons.

Detection

The common means of detecting a charged particle by looking for a track of ionization (such as in a cloud chamber) does not work for neutrons directly. Neutrons that elastically scatter off atoms can create an ionization track that is detectable, but the experiments are not as simple to carry out; other means for detecting neutrons, consisting of allowing them to interact with atomic nuclei, are more commonly used. The commonly used methods to detect neutrons can therefore be categorized according to the nuclear processes relied upon, mainly neutron capture or elastic scattering.

Neutron Detection by Neutron Capture

A common method for detecting neutrons involves converting the energy released from neutron capture reactions into electrical signals. Certain nuclides have a high neutron capture cross section, which is the probability of absorbing a neutron. Upon neutron capture, the compound nucleus emits more easily detectable radiation, for example an alpha particle, which is then detected. The nuclides 3He, 6Li, 10B, 233U, 235U, 237Np and 239Pu are useful for this purpose.

Neutron Detection by Elastic Scattering

Neutrons can elastically scatter off nuclei, causing the struck nucleus to recoil. Kinematically, a neutron can transfer more energy to light nuclei such as hydrogen or helium than to heavier nuclei. Detectors relying on elastic scattering are called fast neutron detectors. Recoiling nuclei can ionize and excite further atoms through collisions. Charge and/or scintillation light produced in this way can be collected to produce a detected signal. A major challenge in fast neutron detection is discerning such signals from erroneous signals produced by gamma radiation in the same detector.

Fast neutron detectors have the advantage of not requiring a moderator, and therefore being capable of measuring the neutron's energy, time of arrival, and in certain cases direction of incidence.

Sources and Production

Free neutrons are unstable, although they have the longest half-life of any unstable sub-atomic particle by several orders of magnitude. Their half-life is still only about 10 minutes, however, so they can be obtained only from sources that produce them freshly.

A small natural background flux of free neutrons exists everywhere on Earth. In the atmosphere and deep into the ocean, the "neutron background" is caused by muons produced by cosmic ray interaction with the atmosphere. These high energy muons are capable of penetration to considerable depths in water and soil. There, in striking atomic nuclei, among other reactions they induce spallation reactions in which a neutron is liberated from the nucleus.

Within the Earth's crust a second source is neutrons produced primarily by spontaneous fission of uranium and thorium present in crustal minerals. The neutron background is not strong enough to be a biological hazard, but it is of importance to very high resolution particle detectors that are looking for very rare events, such as (hypothesized) interactions that might be caused by particles of dark matter. Recent research has shown that even thunderstorms can produce neutrons with energies of up to several tens of MeV.

Even stronger neutron background radiation is produced at the surface of Mars, where the atmosphere is thick enough to generate neutrons from cosmic ray muon production and neutron-spallation, but not thick enough to provide significant protection from the neutrons produced. These neutrons not only produce a Martian surface neutron radiation hazard from direct downward-going neutron radiation but may also produce a significant hazard from reflection of neutrons from the Martian surface, which will produce reflected neutron radiation penetrating upward into a Martian craft or habitat from the floor.

These include certain types of radioactive decay (spontaneous fission and neutron emission), and from certain nuclear reactions. Convenient nuclear reactions include tabletop reactions such as natural alpha and gamma bombardment of certain nuclides, often beryllium or deuterium, and induced nuclear fission, such as occurs in nuclear reactors. In addition, high-energy nuclear reactions (such as occur in cosmic radiation showers or accelerator collisions) also produce neutrons from disintigration of target nuclei. Small (tabletop) particle accelerators optimized to produce free neutrons in this way, are called neutron generators.

In practice, the most commonly used small laboratory sources of neutrons use radioactive decay to power neutron production. One noted neutron-producing radioisotope, californium-252 decays (half-life 2.65 years) by spontaneous fission 3% of the time with production of 3.7 neutrons per fission, and is used alone as a neutron source from this process. Nuclear reaction sources (that involve two materials) powered by radioisotopes use an alpha decay source plus a beryllium target, or else a source of high-energy gamma radiation from a source that undergoes beta decay followed by gamma decay, which produces photoneutrons on interaction of the high energy gamma ray with ordinary stable beryllium, or else with the deuterium in heavy water. A popular source of the latter type is radioactive antimony-124 plus beryllium, a system with a half-life of 60.9 days, which can be constructed from natural antimony (which is 42.8% stable antimony-123) by activating it with neutrons in a nuclear reactor, then transported to where the neutron source is needed.

Institut Laue–Langevin (ILL) in Grenoble, France – a major neutron research facility.

Nuclear fission reactors naturally produce free neutrons; their role is to sustain the energy-producing chain reaction. The intense neutron radiation can also be used to produce various radioisotopes through the process of neutron activation, which is a type of neutron capture.

Experimental nuclear fusion reactors produce free neutrons as a waste product. However, it is these neutrons that possess most of the energy, and converting that energy to a useful form has proved a difficult engineering challenge. Fusion reactors that generate neutrons are likely to create radioactive waste, but the waste is composed of neutron-activated lighter isotopes, which have relatively short (50–100 years) decay periods as compared to typical half-lives of 10,000 years for fission waste, which is long due primarily to the long half-life of alpha-emitting transuranic actinides.

Neutron Beams and Modification of Beams after Production

Free neutron beams are obtained from neutron sources by neutron transport. For access to intense neutron sources, researchers must go to a specialist neutron facility that operates a research reactor or a spallation source.

The neutron's lack of total electric charge makes it difficult to steer or accelerate them. Charged particles can be accelerated, decelerated, or deflected by electric or magnetic fields. These methods have little effect on neutrons. However, some effects may be attained by use of inhomogeneous magnetic fields because of the neutron's magnetic moment. Neutrons can be controlled by methods that include moderation, reflection, and velocity selection. Thermal neutrons can be polarized by transmission through magnetic materials in a method analogous to the Faraday effect for photons. Cold neutrons of wavelengths of 6–7 angstroms can be produced in beams of a high degree of polarization, by use of magnetic mirrors and magnetized interference filters.

Applications

The neutron plays an important role in many nuclear reactions. For example, neutron capture often results in neutron activation, inducing radioactivity. In particular, knowledge of neutrons and their behavior has been important in the development of nuclear reactors and nuclear weapons. The fissioning of elements like uranium-235 and plutonium-239 is caused by their absorption of neutrons.

Cold, thermal and *hot* neutron radiation is commonly employed in neutron scattering facilities, where the radiation is used in a similar way one uses X-rays for the analysis of condensed matter. Neutrons are complementary to the latter in terms of atomic contrasts by different scattering cross sections; sensitivity to magnetism; energy range for inelastic neutron spectroscopy; and deep penetration into matter.

The development of "neutron lenses" based on total internal reflection within hollow glass capillary tubes or by reflection from dimpled aluminum plates has driven ongoing research into neutron microscopy and neutron/gamma ray tomography.

A major use of neutrons is to excite delayed and prompt gamma rays from elements in materials. This forms the basis of neutron activation analysis (NAA) and prompt gamma neutron activation analysis (PGNAA). NAA is most often used to analyze small samples of materials in a nuclear reactor whilst PGNAA is most often used to analyze subterranean rocks around bore holes and industrial bulk materials on conveyor belts.

Another use of neutron emitters is the detection of light nuclei, in particular the hydrogen found in water molecules. When a fast neutron collides with a light nucleus, it loses a large fraction of its energy. By measuring the rate at which slow neutrons return to the probe after reflecting off of hydrogen nuclei, a neutron probe may determine the water content in soil.

Medical Therapies

Because neutron radiation is both penetrating and ionizing, it can be exploited for medical treatments. Neutron radiation can have the unfortunate side-effect of leaving the affected area radioactive, however. Neutron tomography is therefore not a viable medical application.

Fast neutron therapy utilizes high energy neutrons typically greater than 20 MeV to treat cancer. Radiation therapy of cancers is based upon the biological response of cells to ionizing radiation. If radiation is delivered in small sessions to damage cancerous areas, normal tissue will have time to repair itself, while tumor cells often cannot. Neutron radiation can deliver energy to a cancerous region at a rate an order of magnitude larger than gamma radiation

Beams of low energy neutrons are used in boron capture therapy to treat cancer. In boron capture therapy, the patient is given a drug that contains boron and that preferentially accumulates in the tumor to be targeted. The tumor is then bombarded with very low energy neutrons (although often higher than thermal energy) which are captured by the boron-10 isotope in the boron, which produces an excited state of boron-11 that then decays to produce lithium-7 and an alpha particle that have sufficient energy to kill the malignant cell, but insufficient range to damage nearby cells. For such a therapy to be applied to the treatment of cancer, a neutron source having an intensity of the order of billion (10^9) neutrons per second per cm^2 is preferred. Such fluxes require a research nuclear reactor.

Protection

Exposure to free neutrons can be hazardous, since the interaction of neutrons with molecules in the body can cause disruption to molecules and atoms, and can also cause reactions that give rise to other forms of radiation (such as protons). The normal precautions of radiation protection apply: Avoid exposure, stay as far from the source as possible, and keep exposure time to a minimum. Some particular thought must be given to how to protect from neutron exposure, however. For other types of radiation, e.g., alpha particles, beta particles, or gamma rays, material of a high atomic number and with high density make for good shielding; frequently, lead is used. However, this approach will not work with neutrons, since the absorption of neutrons does not increase straightforwardly with atomic number, as it does with alpha, beta, and gamma radiation. Instead one needs to look at the particular interactions neutrons have with matter (see the section on detection above). For example, hydrogen-rich materials are often used to shield against neutrons, since ordinary hydrogen both scatters and slows neutrons. This often means that simple concrete blocks or even paraffin-loaded plastic blocks afford better protection from neutrons than do far more dense materials. After slowing, neutrons may then be absorbed with an isotope that has high affinity for slow neutrons without causing secondary capture radiation, such as lithium-6.

Hydrogen-rich ordinary water affects neutron absorption in nuclear fission reactors: Usually, neutrons are so strongly absorbed by normal water that fuel enrichment with fissionable isotope is required. The deuterium in heavy water has a very much lower absorption affinity for neutrons than does protium (normal light hydrogen). Deuterium is, therefore, used in CANDU-type reactors, in order to slow (moderate) neutron velocity, to increase the probability of nuclear fission compared to neutron capture.

Neutron Temperature

Thermal Neutrons

A *thermal neutron* is a free neutron that is Boltzmann distributed with kT = 0.0253 eV (4.0×10^{-21} J) at room temperature. This gives characteristic (not average, or median) speed of 2.2 km/s. The name 'thermal' comes from their energy being that of the room temperature gas or material they are permeating. After a number of colli-sions (often in the range of 10–20) with nuclei, neutrons arrive at this energy level, provided that they are not absorbed.

In many substances, thermal neutron reactions show a much larger effective cross-section than reactions involving faster neutrons, and thermal neutrons can therefore be absorbed more readily (i.e., with higher probability) by any atomic nuclei that they collide with, creating a heavier — and often unstable — isotope of the chemical element as a result.

Most fission reactors use a neutron moderator to slow down, or *thermalize* the neutrons that are emitted by nuclear fission so that they are more easily captured, causing further fission. Others, called fast breeder reactors, use fission energy neutrons directly.

Cold Neutrons

Cold neutrons are thermal neutrons that have been equilibrated in a very cold substance such as liquid deuterium. Such a *cold source* is placed in the moderator of a research reactor or spallation source. Cold neutrons are particularly valuable for neutron scattering experiments.

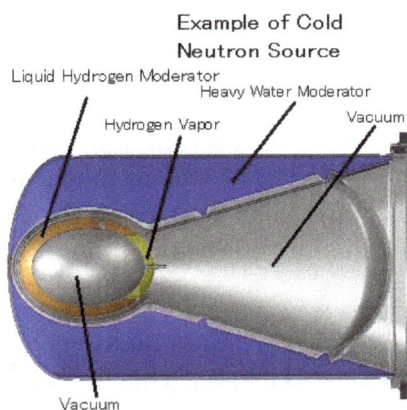

Cold neutron source providing neutrons at about the temperature of liquid hydrogen

Ultracold Neutrons

Ultracold neutrons are produced by inelastically scattering cold neutrons in substances with a temperature of a few kelvins, such as solid deuterium or superfluid helium. An alternative produc-tion method is the mechanical deceleration of cold neutrons.

Fission Energy Neutrons

A *fast neutron* is a free neutron with a kinetic energy level close to 1 MeV (1.6×10^{-13} J), hence a speed of ~14000 km/s (~ 5% of the speed of light). They are named *fission energy* or *fast* neutrons to distinguish them from lower-energy thermal neutrons, and high-energy neutrons produced in cosmic showers or accelerators. Fast neutrons are produced by nuclear processes such as nuclear fission. Neutrons produced in fission, as noted above, have a Maxwell–Boltzmann distribution of kinetic energies from 0 to ~14 MeV, a mean energy of 2 MeV (for U-235 fission neutrons), and a mode of only 0.75 MeV, which means that more than half of them do not qualify as fast (and thus have almost no chance of initiating fission in fertile materials, such as U-238 and Th-232).

Fast neutrons can be made into thermal neutrons via a process called moderation. This is done with a neutron moderator. In reactors, typically heavy water, light water, or graphite are used to moderate neutrons.

Fusion Neutrons

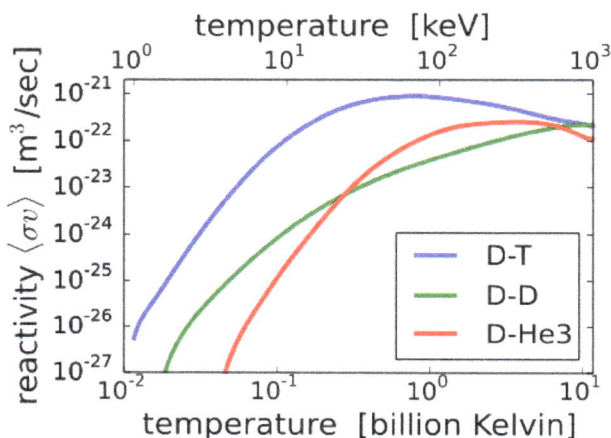

The fusion reaction rate increases rapidly with temperature until it maximizes and then gradually drops off. The DT rate peaks at a lower temperature (about 70 keV, or 800 million kelvins) and at a higher value than other reactions commonly considered for fusion energy.

D–T (deuterium–tritium) fusion is the fusion reaction that produces the most energetic neutrons, with 14.1 MeV of kinetic energy and traveling at 17% of the speed of light. D–T fusion is also the easiest fusion reaction to ignite, reaching near-peak rates even when the deuterium and tritium nuclei have only a thousandth as much kinetic energy as the 14.1 MeV that will be produced.

14.1 MeV neutrons have about 10 times as much energy as fission neutrons, and are very effective at fissioning even non-fissile heavy nuclei, and these high-energy fissions produce more neutrons on average than fissions by lower-energy neutrons. This makes D–T fusion neutron sources such as proposed tokamak power reactors useful for transmutation of transuranic waste. 14.1 MeV neutrons can also produce neutrons by knocking them loose from nuclei.

On the other hand, these very high energy neutrons are less likely to simply be captured without causing fission or spallation. For these reasons, nuclear weapon design extensively utilizes D–T fusion 14.1 MeV neutrons to cause more fission. Fusion neutrons are able to cause fission in ordinarily non-fissile materials, such as depleted uranium (uranium-238), and these materials have been used in the jackets of thermonuclear weapons. Fusion neutrons also can cause fission in substances that are unsuitable or difficult to make into primary fission bombs, such as reactor grade plutonium. This physical fact thus causes ordinary non-weapons grade materials to become of concern in certain nuclear proliferation discussions and treaties.

Other fusion reactions produce much less energetic neutrons. D–D fusion produces a 2.45 MeV neutron and helium-3 half of the time, and produces tritium and a proton but no neutron the other half of the time. D–^{3}He fusion produces no neutron.

Intermediate-Energy Neutrons

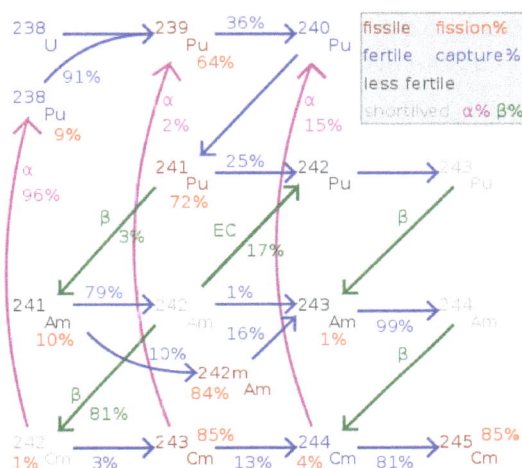

Transmutation flow in light water reactor, which is a thermal-spectrum reactor

A fission energy neutron that has slowed down but not yet reached thermal energies is called an epithermal neutron.

Cross sections for both capture and fission reactions often have multiple resonance peaks at specific energies in the epithermal energy range. These are of less significance in a fast neutron reactor, where most neutrons are absorbed before slowing down to this range, or in a well-moderated thermal reactor, where epithermal neutrons interact mostly with moderator nuclei, not with either fissile or fertile actinide nuclides. However, in a partially moderated reactor with more interactions of epithermal neutrons with heavy metal nuclei, there are greater possibilities for transient changes in reactivity that might make reactor control more difficult.

Ratios of capture reactions to fission reactions are also worse (more captures without fission) in most nuclear fuels such as plutonium-239, making epithermal-spectrum reactors using these fuels less desirable, as captures not only waste the one neutron captured but also usually result in a nuclide that is not fissile with thermal or epithermal neutrons, though still fissionable with fast neutrons. The exception is uranium-233 of the thorium cycle, which has good capture-fission ratios at all neutron energies.

High-Energy Neutrons

These neutrons have much more energy than fission energy neutrons and are generated as secondary particles by particle accelerators or in the atmosphere from cosmic rays. These high-energy neutrons are extremely efficient at ionization and far more likely to cause cell death than X-rays or protons.

Chemical Element

A chemical element or element is a species of atoms having the same number of protons in their atomic nuclei (i.e. the same atomic number, Z). There are 118 elements that have been identified, of which the first 94 occur naturally on Earth with the remaining 24 being synthetic elements. There are 80 elements that have at least one stable isotope and 38 that have exclusively radioactive isotopes, which decay over time into other elements. Iron is the most abundant element (by mass) making up Earth, while oxygen is the most common element in the crust of Earth.

Chemical elements constitute all of the ordinary matter of the universe. However astronomical observations suggest that ordinary observable matter is only approximately 15% of the matter in the universe: the remainder is dark matter, the composition of which is unknown, but it is not composed of chemical elements. The two lightest elements, hydrogen and helium were mostly formed in the Big Bang and are the most common elements in the universe. The next three elements (lithium, beryllium and boron) were formed mostly by cosmic ray spallation, and are thus more rare than those that follow. Formation of elements with from six to twenty six protons occurred and continues to occur in main sequence stars via stellar nucleosynthesis. The high abundance of oxygen, silicon, and iron on Earth reflects their common production in such stars. Elements with greater than twenty-six protons are formed by supernova nucleosynthesis in supernovae, which, when they explode, blast these elements far into space as supernova remnants, where they may become incorporated into planets when they are formed.

The term "element" is used for a kind of atoms with a given number of protons (regardless of whether they are or they are not ionized or chemically bonded, e.g. hydrogen in water) as well as for a pure chemical substance consisting of a single element (e.g. hydrogen gas). For the second meaning, the terms "elementary substance" and "simple substance" have been suggested, but they have not gained much acceptance in the English-language chemical literature, whereas in some other languages their equivalent is widely used (e.g. French *corps simple*, Russian *простое вещество*). One element can form multiple substances different by their structure; they are called allotropes of the element.

When different elements are chemically combined, with the atoms held together by chemical bonds, they form chemical compounds. Only a minority of elements are found uncombined as relatively pure minerals. Among the more common of such "native elements" are copper, silver, gold, carbon (as coal, graphite, or diamonds), and sulfur. All but a few of the most inert elements, such as noble gases and noble metals, are usually found on Earth in chemically combined form, as chemical compounds. While about 32 of the chemical elements occur on Earth in native uncombined forms, most of these occur as mixtures. For example, atmospheric air is primarily a mixture of nitrogen, oxygen, and argon, and native solid elements occur in alloys, such as that of iron and nickel.

The history of the discovery and use of the elements began with primitive human societies that found native elements like carbon, sulfur, copper and gold. Later civilizations extracted elemental copper, tin, lead and iron from their ores by smelting, using charcoal. Alchemists and chemists subsequently identified many more, with almost all of the naturally-occurring elements becoming known by 1900.

The properties of the chemical elements are summarized on the periodic table, which organizes the elements by increasing atomic number into rows ("periods") in which the columns ("groups") share recurring ("periodic") physical and chemical properties. Save for unstable radioactive elements with short half-lives, all of the elements are available industrially, most of them in high degrees of purity.

Description

The lightest chemical elements are hydrogen and helium, both created by Big Bang nucleosynthesis during the first 20 minutes of the universe in a ratio of around 3:1 by mass (or 12:1 by number of atoms), along with tiny traces of the next two elements, lithium and beryllium. Almost all other elements found in nature were made by various natural methods of nucleosynthesis. On Earth, small amounts of new atoms are naturally produced in nucleogenic reactions, or in cosmogenic processes, such as cosmic ray spallation. New atoms are also naturally produced on Earth as radiogenic daughter isotopes of ongoing radioactive decay processes such as alpha decay, beta decay, spontaneous fission, cluster decay, and other rarer modes of decay.

Of the 94 naturally occurring elements, those with atomic numbers 1 through 82 each have at least one stable isotope (except for technetium, element 43 and promethium, element 61, which have no stable isotopes). Isotopes considered stable are those for which no radioactive decay has yet been observed. Elements with atomic numbers 83 through 94 are unstable to the point that radioactive decay of all isotopes can be detected. Some of these elements, notably bismuth (atomic number 83), thorium (atomic number 90), uranium (atomic number 92) and plutonium (atomic number 94), have one or more isotopes with half-lives long enough to survive as remnants of the explosive stellar nucleosynthesis that produced the heavy elements before the formation of our solar system. At over 1.9×10^{19} years, over a billion times longer than the current estimated age of the universe, bismuth-209 (atomic number 83) has the longest known alpha decay half-life of any naturally occurring element, and is almost always considered on par with the 80 stable elements. The very heaviest elements (those beyond plutonium, element 94) undergo radioactive decay with half-lives so short that they are not found in nature and must be synthesized.

As of 2010, there are 118 known elements (in this context, "known" means observed well enough, even from just a few decay products, to have been differentiated from other elements). Of these 118 elements, 94 occur naturally on Earth. Six of these occur in extreme trace quantities: technetium, atomic number 43; promethium, number 61; astatine, number 85; francium, number 87; neptunium, number 93; and plutonium, number 94. These 94 elements have been detected in the universe at large, in the spectra of stars and also supernovae, where short-lived radioactive elements are newly being made. The first 94 elements have been detected directly on Earth as primordial nuclides present from the formation of the solar system, or as naturally-occurring fission or transmutation products of uranium and thorium.

The remaining 24 heavier elements, not found today either on Earth or in astronomical spectra, have been produced artificially: these are all radioactive, with very short half-lives; if any atoms of these elements were present at the formation of Earth, they are extremely likely, to the point of certainty, to have already decayed, and if present in novae, have been in quantities too small to have been noted. Technetium was the first purportedly non-naturally occurring element synthesized, in 1937, although trace amounts of technetium have since been found in nature (and also the element may have been discovered naturally in 1925). This pattern of artificial production and later natural discovery has been repeated with several other radioactive naturally-occurring rare elements.

Lists of the elements are available by name, by symbol, by atomic number, by density, by melting point, and by boiling point as well as ionization energies of the elements. The nuclides of stable and radioactive elements are also available as a list of nuclides, sorted by length of half-life for those that are unstable. One of the most convenient, and certainly the most traditional presentation of the elements, is in the form of the periodic table, which groups together elements with similar chemical properties (and usually also similar electronic structures).

Atomic Number

The atomic number of an element is equal to the number of protons in each atom, and defines the element. For example, all carbon atoms contain 6 protons in their atomic nucleus; so the atomic number of carbon is 6. Carbon atoms may have different numbers of neutrons; atoms of the same element having different numbers of neutrons are known as isotopes of the element.

The number of protons in the atomic nucleus also determines its electric charge, which in turn determines the number of electrons of the atom in its non-ionized state. The electrons are placed into atomic orbitals that determine the atom's various chemical properties. The number of neutrons in a nucleus usually has very little effect on an element's chemical properties (except in the case of hydrogen and deuterium). Thus, all carbon isotopes have nearly identical chemical properties because they all have six protons and six electrons, even though carbon atoms may, for example, have 6 or 8 neutrons. That is why the atomic number, rather than mass number or atomic weight, is considered the identifying characteristic of a chemical element.

The symbol for atomic number is Z.

Isotopes

Isotopes are atoms of the same element (that is, with the same number of protons in their atomic nucleus), but having *different* numbers of neutrons. Most (66 of 94) naturally occurring elements have more than one stable isotope. Thus, for example, there are three main isotopes of carbon. All carbon atoms have 6 protons in the nucleus, but they can have either 6, 7, or 8 neutrons. Since the mass numbers of these are 12, 13 and 14 respectively, the three isotopes of carbon are known as carbon-12, carbon-13, and carbon-14, often abbreviated to ^{12}C, ^{13}C, and ^{14}C. Carbon in everyday life and in chemistry is a mixture of ^{12}C (about 98.9%), ^{13}C (about 1.1%) and about 1 atom per trillion of ^{14}C.

Except in the case of the isotopes of hydrogen (which differ greatly from each other in relative mass—enough to cause chemical effects), the isotopes of a given element are chemically nearly indistinguishable.

All of the elements have some isotopes that are radioactive (radioisotopes), although not all of these radioisotopes occur naturally. The radioisotopes typically decay into other elements upon radiating an alpha or beta particle. If an element has isotopes that are not radioactive, these are termed "stable" isotopes. All of the known stable isotopes occur naturally. The many radioisotopes that are not found in nature have been characterized after being artificially made. Certain elements have no stable isotopes and are composed *only* of radioactive isotopes: specifically the elements without any stable isotopes are technetium (atomic number 43), prome- thium (atomic number 61), and all observed elements with atomic numbers greater than 82.

Of the 80 elements with at least one stable isotope, 26 have only one single stable isotope. The mean number of stable isotopes for the 80 stable elements is 3.1 stable isotopes per element. The largest number of stable isotopes that occur for a single element is 10 (for tin, element 50).

Isotopic Mass and Atomic Mass

The mass number of an element, *A*, is the number of nucleons (protons and neutrons) in the atomic nucleus. Different isotopes of a given element are distinguished by their mass numbers, which are conventionally written as a superscript on the left hand side of the atomic symbol (e.g., ^{238}U). The mass number is always a simple whole number and has units of "nucleons." An example of a referral to a mass number is "magnesium-24," which is an atom with 24 nucleons (12 protons and 12 neutrons).

Whereas the mass number simply counts the total number of neutrons and protons and is thus a natural (or whole) number, the atomic mass of a single atom is a real number for the mass of a particular isotope of the element, the unit being **u**. In general, when expressed in **u** it differs in value slightly from the mass number for a given nuclide (or isotope) since the mass of the protons and neutrons is not exactly 1 **u**, since the electrons contribute a lesser share to the atomic mass as neutron number exceeds proton number, and (finally) because of the nuclear binding energy. For example, the atomic mass of chlorine-35 to five significant digits is 34.969 **u** and that of chlo- rine-37 is 36.966 **u**. However, the atomic mass in **u** of each isotope is quite close to its simple mass number (always within 1%). The only isotope whose atomic mass is exactly a natural number is ^{12}C, which by definition has a mass of exactly 12, because **u** is defined as 1/12 of the mass of a free neutral carbon-12 atom in the ground state.

The relative atomic mass (historically and commonly also called "atomic weight") of an element is the *average* of the atomic masses of all the chemical element's isotopes as found in a particular environment, weighted by isotopic abundance, relative to the atomic mass unit (**u**). This number may be a fraction that is *not* close to a whole number, due to the averaging process. For example, the relative atomic mass of chlorine is 35.453 **u**, which differs greatly from a whole number due to being made of an average of 76% chlorine-35 and 24% chlorine-37. Whenever a relative atomic mass value differs by more than 1% from a whole number, it is due to this averaging effect resulting from significant amounts of more than one isotope being naturally present in the sample of the element in question.

Chemically Pure and Isotopically Pure

Chemists and nuclear scientists have different definitions of a *pure element*. In chemistry, a pure element means a substance whose atoms all (or in practice almost all) have the same atomic num- ber, or number of protons. Nuclear scientists, however, define a pure element as one that consists of only one stable isotope.

For example, a copper wire is 99.99% chemically pure if 99.99% of its atoms are copper, with 29 protons each. However it is not isotopically pure since ordinary copper consists of two stable isotopes, 69% ^{63}Cu and 31% ^{65}Cu, with different numbers of neutrons. However, a pure gold ingot would be both chemically and isotopically pure, since ordinary gold consists only of one isotope, Au.

Allotropes

Atoms of chemically pure elements may bond to each other chemically in more than one way, allowing the pure element to exist in multiple structures (spatial arrangements of atoms), known as allotropes, which differ in their properties. For example, carbon can be found as diamond, which has a tetrahedral structure around each carbon atom; graphite, which has layers of carbon atoms with a hexagonal structure stacked on top of each other; graphene, which is a single layer of graphite that is very strong; fullerenes, which have nearly spherical shapes; and carbon nanotubes, which are tubes with a hexagonal structure (even these may differ from each other in electrical properties). The ability of an element to exist in one of many structural forms is known as 'allotropy'.

The standard state, also known as reference state, of an element is defined as its thermodynamically most stable state at 1 bar at a given temperature (typically at 298.15 K). In thermochemistry, an element is defined to have an enthalpy of formation of zero in its standard state. For example, the reference state for carbon is graphite, because the structure of graphite is more stable than that of the other allotropes.

Properties

Several kinds of descriptive categorizations can be applied broadly to the elements, including consideration of their general physical and chemical properties, their states of matter under familiar conditions, their melting and boiling points, their densities, their crystal structures as solids, and their origins.

General Properties

Several terms are commonly used to characterize the general physical and chemical properties of the chemical elements. A first distinction is between metals, which readily conduct electricity, nonmetals, which do not, and a small group, (the *metalloids*), having intermediate properties and often behaving as semiconductors.

A more refined classification is often shown in colored presentations of the periodic table. This system restricts the terms "metal" and "nonmetal" to only certain of the more broadly defined metals and nonmetals, adding additional terms for certain sets of the more broadly viewed metals and nonmetals. The version of this classification used in the periodic tables presented here includes: actinides, alkali metals, alkaline earth metals, halogens, lanthanides, transition metals, post-transition metals, metalloids, polyatomic nonmetals, diatomic nonmetals, and noble gases. In this system, the alkali metals, alkaline earth metals, and transition metals, as well as the lanthanides and the actinides, are special groups of the metals viewed in a broader sense. Similarly, the polyatomic nonmetals, diatomic nonmetals and the noble gases are nonmetals viewed in the broader sense. In some presentations, the halogens are not distinguished, with astatine identified as a metalloid and the others identified as nonmetals.

States of Matter

Another commonly used basic distinction among the elements is their state of matter (phase), whether solid, liquid, or gas, at a selected standard temperature and pressure (STP). Most of the elements are solids at conventional temperatures and atmospheric pressure, while several are gases. Only bromine and mercury are liquids at 0 degrees Celsius (32 degrees Fahrenheit) and normal atmospheric pressure; caesium and gallium are solids at that temperature, but melt at 28.4 °C (83.2 °F) and 29.8 °C (85.6 °F), respectively.

Melting and Boiling Points

Melting and boiling points, typically expressed in degrees Celsius at a pressure of one atmosphere, are commonly used in characterizing the various elements. While known for most elements, either or both of these measurements is still undetermined for some of the radioactive elements available in only tiny quantities. Since helium remains a liquid even at absolute zero at atmospheric pressure, it has only a boiling point, and not a melting point, in conventional presentations.

Densities

The density at a selected standard temperature and pressure (STP) is frequently used in characterizing the elements. Density is often expressed in grams per cubic centimeter (g/cm³). Since several elements are gases at commonly encountered temperatures, their densities are usually stated for their gaseous forms; when liquefied or solidified, the gaseous elements have densities similar to those of the other elements.

When an element has allotropes with different densities, one representative allotrope is typically selected in summary presentations, while densities for each allotrope can be stated where more detail is provided. For example, the three familiar allotropes of carbon (amorphous carbon, graphite, and diamond) have densities of 1.8–2.1, 2.267, and 3.515 g/cm³, respectively.

Crystal Structures

The elements studied to date as solid samples have eight kinds of crystal structures: cubic, body-centered cubic, face-centered cubic, hexagonal, monoclinic, orthorhombic, rhombohedral, and tetragonal. For some of the synthetically produced transuranic elements, available samples have been too small to determine crystal structures.

Occurrence and Origin on Earth

Chemical elements may also be categorized by their origin on Earth, with the first 94 considered naturally occurring, while those with atomic numbers beyond 94 have only been produced artificially as the synthetic products of man-made nuclear reactions.

Of the 94 naturally occurring elements, 84 are considered primordial and either stable or weakly radioactive. The remaining 10 naturally occurring elements possess half lives too short for them to have been present at the beginning of the Solar System, and are therefore considered transient elements. (Plutonium is usually also considered a transient element because primordial plutonium has by now decayed to almost undetectable traces.) Of these 10 transient elements, 5 (polonium, radon, radium, actinium, and protactinium) are relatively common decay products of thorium, uranium, and plutonium. The remaining 5 transient elements (technetium, promethium, astatine, francium, and neptunium) occur only rarely, as products of rare decay modes or nuclear reaction processes involving uranium or other heavy elements, and almost all natural plutonium occurs as such a product rather than as primordial plutonium-244 (half-life 80 million years).

Elements with atomic numbers 1 through 40 are all stable, while those with atomic numbers 41 through 82 (except technetium and promethium) are metastable. The half-lives of these metastable "theoretical radionuclides" are so long (at least 100 million times longer than the estimated age of the universe) that their radioactive decay has yet to be detected by experiment. Elements with atomic numbers 83 through 94 are unstable to the point that their radioactive decay can be detected. Four of these elements, bismuth (element 83), thorium (element 90), uranium (element 92), and plutonium (element 94), have one or more isotopes with half-lives long enough to survive as remnants of the explosive stellar nucleosynthesis that produced the heavy elements before the formation of our solar system. For example, at over 1.9×10^{19} years, over a billion times longer than the current estimated age of the universe, bismuth-209 has the longest known alpha decay half-life of any naturally occurring element. The very heaviest 24 elements (those beyond plutonium, element 94) undergo radioactive decay with short half-lives and cannot be produced as daughters of longer-lived elements, and thus they do not occur in nature at all.

The Periodic Table

The properties of the chemical elements are often summarized using the periodic table, which powerfully and elegantly organizes the elements by increasing atomic number into rows ("periods") in which the columns ("groups") share recurring ("periodic") physical and chemical properties. The current standard table contains 118 confirmed elements as of 10 April 2010.

Although earlier precursors to this presentation exist, its invention is generally credited to the Russian chemist Dmitri Mendeleev in 1869, who intended the table to illustrate recurring trends in the properties of the elements. The layout of the table has been refined and extended over time as new elements have been discovered and new theoretical models have been developed to explain chemical behavior.

Use of the periodic table is now ubiquitous within the academic discipline of chemistry, providing an extremely useful framework to classify, systematize and compare all the many different forms of chemical behavior. The table has also found wide application in physics, geology, biology, materials science, engineering, agriculture, medicine, nutrition, environmental health, and astronomy. Its principles are especially important in chemical engineering.

Nomenclature and Symbols

The various chemical elements are formally identified by their unique atomic numbers, by their accepted names, and by their symbols.

Atomic Numbers

The known elements have atomic numbers from 1 through 118, conventionally presented as Arabic numerals. Since the elements can be uniquely sequenced by atomic number, conventionally from lowest to highest (as in a periodic table), sets of elements are sometimes specified by such notation as "through", "beyond", or "from ... through", as in "through iron", "beyond uranium", or "from lanthanum through lutetium". The terms "light" and "heavy" are sometimes also used informally to indicate relative atomic numbers (not densities), as in "lighter than carbon" or "heavier than lead", although technically the weight or mass of atoms of an element (their atomic weights or atomic masses) do not always increase monotonically with their atomic numbers.

Element Names

The naming of various substances now known as elements precedes the atomic theory of matter, as names were given locally by various cultures to various minerals, metals, compounds, alloys, mixtures, and other materials, although at the time it was not known which chemicals were elements and which compounds. As they were identified as elements, the existing names for anciently-known elements (e.g., gold, mercury, iron) were kept in most countries. National differences emerged over the names of elements either for convenience, linguistic niceties, or nationalism. For a few illustrative examples: German speakers use "Wasserstoff" (water substance) for "hydrogen", "Sauerstoff" (acid substance) for "oxygen" and "Stickstoff" (smothering substance) for "nitrogen", while English and some romance languages use "sodium" for "natrium" and "potassium" for "kalium", and the French, Italians, Greeks, Portuguese and Poles prefer "azote/azot/azoto" (from roots meaning "no life") for "nitrogen".

For purposes of international communication and trade, the official names of the chemical elements both ancient and more recently recognized are decided by the International Union of Pure and Applied Chemistry (IUPAC), which has decided on a sort of international English language, drawing on traditional English names even when an element's chemical symbol is based on a Latin or other traditional word, for example adopting "gold" rather than "aurum" as the name for the 79th element (Au). IUPAC prefers the British spellings "aluminium" and "caesium" over the U.S. spellings "aluminum" and "cesium", and the U.S. "sulfur" over the British "sulphur". However, elements that are practical to sell in bulk in many countries often still have locally used national names, and countries whose national language does not use the Latin alphabet are likely to use the IUPAC element names.

According to IUPAC, chemical elements are not proper nouns in English; consequently, the full name of an element is not routinely capitalized in English, even if derived from a proper noun, as in californium and einsteinium. Isotope names of chemical elements are also uncapitalized if written out, *e.g.*, carbon-12 or uranium-235. Chemical element *symbols* (such as Cf for californium and Es for einsteinium), are always capitalized.

In the second half of the twentieth century, physics laboratories became able to produce nuclei of chemical elements with half-lives too short for an appreciable amount of them to exist at any time. These are also named by IUPAC, which generally adopts the name chosen by the discoverer. This practice can lead to the controversial question of which research group actually discovered an element, a question that delayed the naming of elements with atomic number of 104 and higher for a considerable amount of time.

Precursors of such controversies involved the nationalistic namings of elements in the late 19th century. For example, *lutetium* was named in reference to Paris, France. The Germans were reluctant to relinquish naming rights to the French, often calling it *cassiopeium*. Similarly, the British discoverer of *niobium* originally named it *columbium,* in reference to the New World. It was used extensively as such by American publications prior to the international standardization (in 1950).

Chemical Symbols

For listings of current chemical symbols, symbols not currently used, and other symbols that may look like chemical symbols, see Symbol (chemistry).

Specific Chemical Elements

Before chemistry became a science, alchemists had designed arcane symbols for both metals and common compounds. These were however used as abbreviations in diagrams or procedures; there was no concept of atoms combining to form molecules. With his advances in the atomic theory of matter, John Dalton devised his own simpler symbols, based on circles, to depict molecules.

The current system of chemical notation was invented by Berzelius. In this typographical system, chemical symbols are not mere abbreviations—though each consists of letters of the Latin alphabet. They are intended as universal symbols for people of all languages and alphabets.

The first of these symbols were intended to be fully universal. Since Latin was the common language of science at that time, they were abbreviations based on the Latin names of metals. Cu comes from Cuprum, Fe comes from Ferrum, Ag from Argentum. The symbols were not followed by a period (full stop) as with abbreviations. Later chemical elements were also assigned unique chemical symbols, based on the name of the element, but not necessarily in English. For example, sodium has the chemical symbol 'Na' after the Latin *natrium*. The same applies to "W" (wolfram) for tungsten, "Fe" (ferrum) for iron, "Hg" (hydrargyrum) for mercury, "Sn" (stannum) for tin, "K" (kalium) for potassium, "Au" (aurum) for gold, "Ag" (argentum) for silver, "Pb" (plumbum) for lead, "Cu" (cuprum) for copper, and "Sb" (stibium) for antimony.

Chemical symbols are understood internationally when element names might require translation. There have sometimes been differences in the past. For example, Germans in the past have used "J" (for the alternate name Jod) for iodine, but now use "I" and "Iod".

The first letter of a chemical symbol is always capitalized, as in the preceding examples, and the subsequent letters, if any, are always lower case (small letters). Thus, the symbols for californium or einsteinium are Cf and Es.

General Chemical Symbols

There are also symbols in chemical equations for groups of chemical elements, for example in comparative formulas. These are often a single capital letter, and the letters are reserved and not used for names of specific elements. For example, an "X" indicates a variable group (usually a halogen) in a class of compounds, while "R" is a radical, meaning a compound structure such as a hydrocarbon chain. The letter "Q" is reserved for "heat" in a chemical reaction. "Y" is also often used as a general chemical symbol, although it is also the symbol of yttrium. "Z" is also frequently used as a general variable group. "E" is used in organic chemistry to denote an electron-withdrawing group or an electrophile; similarly "Nu" denotes a nucleophile. "L" is used to represent a general ligand in inorganic and organometallic chemistry. "M" is also often used in place of a general metal.

At least two additional, two-letter generic chemical symbols are also in informal usage, "Ln" for any lanthanide element and "An" for any actinide element. "Rg" was formerly used for any rare gas element, but the group of rare gases has now been renamed noble gases and the symbol "Rg" has now been assigned to the element roentgenium.

Isotope Symbols

Isotopes are distinguished by the atomic mass number (total protons and neutrons) for a particular isotope of an element, with this number combined with the pertinent element's symbol. IUPAC prefers that isotope symbols be written in superscript notation when practical, for example ^{12}C and ^{235}U. However, other notations, such as carbon-12 and uranium-235, or C-12 and U-235, are also used.

As a special case, the three naturally occurring isotopes of the element hydrogen are often specified as **H** for ^1H (protium), **D** for ^2H (deuterium), and **T** for ^3H (tritium). This convention is easier to use in chemical equations, replacing the need to write out the mass number for each atom. For example, the formula for heavy water may be written D_2O instead of ^2H_2O.

Origin of the Elements

Only about 4% of the total mass of the universe is made of atoms or ions, and thus represented by chemical elements. This fraction is about 15% of the total matter, with the remainder of the matter (85%) being dark matter. The nature of dark matter is unknown, but it is not composed of atoms of chemical elements because it contains no protons, neutrons, or electrons. (The remaining non-matter part of the mass of the universe is composed of the even more mysterious dark energy).

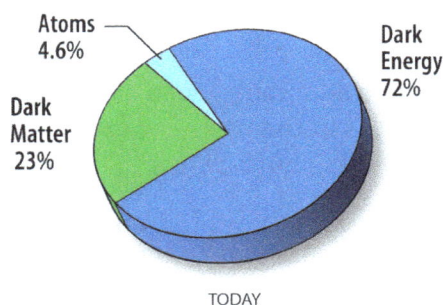

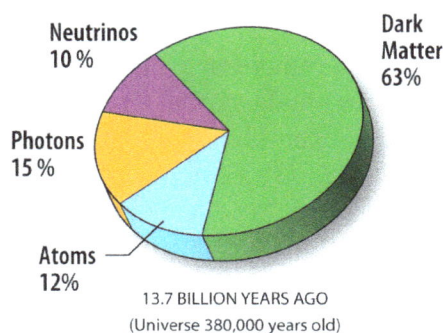

Estimated distribution of dark matter and dark energy in the universe. Only the fraction of the mass and energy in the universe labeled "atoms" is composed of chemical elements.

The universe's 94 naturally occurring chemical elements are thought to have been produced by at least four cosmic processes. Most of the hydrogen and helium in the universe was produced primordially in the first few minutes of the Big Bang. Three recurrently occurring later processes are thought to have produced the remaining elements. Stellar nucleosynthesis, an ongoing process, produces all elements from carbon through iron in atomic number, but little lithium, beryllium, or boron. Elements heavier in atomic number than iron, as heavy as uranium and plutonium, are produced by explosive nucleosynthesis in supernovas and other cataclysmic cosmic events. Cosmic ray spallation (fragmentation) of carbon, nitrogen, and oxygen is important to the production of lithium, beryllium and boron.

During the early phases of the Big Bang, nucleosynthesis of hydrogen nuclei resulted in the production of hydrogen-1 (protium, ^{1}H) and helium-4 (^{4}He), as well as a smaller amount of deuterium (^{2}H) and very minuscule amounts (on the order of 10^{-10}) of lithium and beryllium. Even smaller amounts of boron may have been produced in the Big Bang, since it has been observed in some very old stars, while carbon has not. It is generally agreed that no heavier elements than boron were produced in the Big Bang. As a result, the primordial abundance of atoms (or ions) consisted of roughly 75% ^{1}H, 25% ^{4}He, and 0.01% deuterium, with only tiny traces of lithium, beryllium, and perhaps boron. Subsequent enrichment of galactic halos occurred due to stellar nucleosynthesis and supernova nucleosynthesis. However, the element abundance in intergalactic space can still closely resemble primordial conditions, unless it has been enriched by some means.

Legend:

- B — Big Bang
- C — Cosmic rays
- L — Large stars
- S — Small stars
- $ — Super-novae
- M — Man-made

Periodic table showing the cosmogenic origin of each element in the Big Bang, or in large or small stars. Small stars can produce certain elements up to sulfur, by the alpha process. Supernovae are needed to produce "heavy" elements (those beyond iron and nickel) rapidly by neutron buildup, in the r-process. Certain large stars slowly produce other elements heavier than iron, in the s-process; these may then be blown into space in the off-gassing of planetary nebulae

On Earth (and elsewhere), trace amounts of various elements continue to be produced from other elements as products of natural transmutation processes. These include some produced by cosmic rays or other nuclear reactions, and others produced as decay products of long-lived primordial nuclides. For example, trace (but detectable) amounts of carbon-14 (^{14}C) are continually produced in the atmosphere by cosmic rays impacting nitrogen atoms, and argon-40 (^{40}Ar) is continually produced by the decay of primordially occurring but unstable potassium-40 (^{40}K). Also, three primordially occurring but radioactive actinides, thorium, uranium, and plutonium, decay through a series of recurrently produced but unstable radioactive elements such as radium and radon, which are transiently present in any sample of these metals or their ores or compounds. Three other radioactive elements, technetium, promethium, and neptu-nium, occur only incidentally in natural materials, produced as individual atoms by natural fission of the nuclei of various heavy elements or in other rare nuclear processes.

Human technology has produced various additional elements beyond these first 94, with those through atomic number 118 now known.

Abundance

The following graph (note log scale) shows the abundance of elements in our solar system. The table shows the twelve most common elements in our galaxy (estimated spectroscopically), as measured in parts per million, by mass. Nearby galaxies that have evolved along similar lines have a corresponding enrichment of elements heavier than hydrogen and helium. The more distant galaxies are being viewed as they appeared in the past, so their abundances of elements appear closer to the primordial mixture. As physical laws and processes appear common throughout the visible universe, however, scientist expect that these galaxies evolved elements in similar abundance.

The abundance of elements in the Solar System is in keeping with their origin from nucleosynthesis in the Big Bang and a number of progenitor supernova stars. Very abundant hydrogen and helium are products of the Big Bang, but the next three elements are rare since they had little time to form in the Big Bang and are not made in stars (they are, however, produced in small quantities by the breakup of heavier elements in interstellar dust, as a result of impact by cosmic rays). Beginning with carbon, elements are produced in stars by buildup from alpha particles (helium nuclei), resulting in an alternatingly larger abundance of elements with even atomic numbers (these are also more stable). In general, such elements up to iron are made in large stars in the process of becoming supernovas. Iron-56 is particularly common, since it is the most stable element that can easily be made from alpha particles (being a product of decay of radioactive nickel-56, ultimately made from 14 helium nuclei). Elements heavier than iron are made in energy-absorbing processes in large stars, and their abundance in the universe (and on Earth) generally decreases with their atomic number.

The abundance of the chemical elements on Earth varies from air to crust to ocean, and in various types of life. The abundance of elements in Earth's crust differs from that in the Solar system (as seen in the Sun and heavy planets like Jupiter) mainly in selective loss of the very lightest elements (hydrogen and helium) and also volatile neon, carbon (as hydrocarbons), nitrogen and sulfur, as a result of solar heating in the early formation of the solar system. Oxygen, the most abundant Earth element by mass, is retained on Earth by combination with silicon. Aluminum at 8% by mass is more common in the Earth's crust than in the universe and solar system, but the composition of the far more bulky mantle, which has magnesium and iron in place of aluminum (which occurs there only at 2% of mass) more closely mirrors the elemental composition of the solar system, save for the noted loss of volatile elements to space, and loss of iron which has migrated to the Earth's core.

The composition of the human body, by contrast, more closely follows the composition of seawater—save that the human body has additional stores of carbon and nitrogen necessary to form the proteins and nucleic acids, together with phosphorus in the nucleic acids and energy transfer molecule adenosine triphosphate (ATP) that occurs in the cells of all living organisms. Certain kinds of organisms require particular additional elements, for example the magnesium in chlorophyll in green plants, the calcium in mollusc shells, or the iron in the hemoglobin in vertebrate animals' red blood cells.

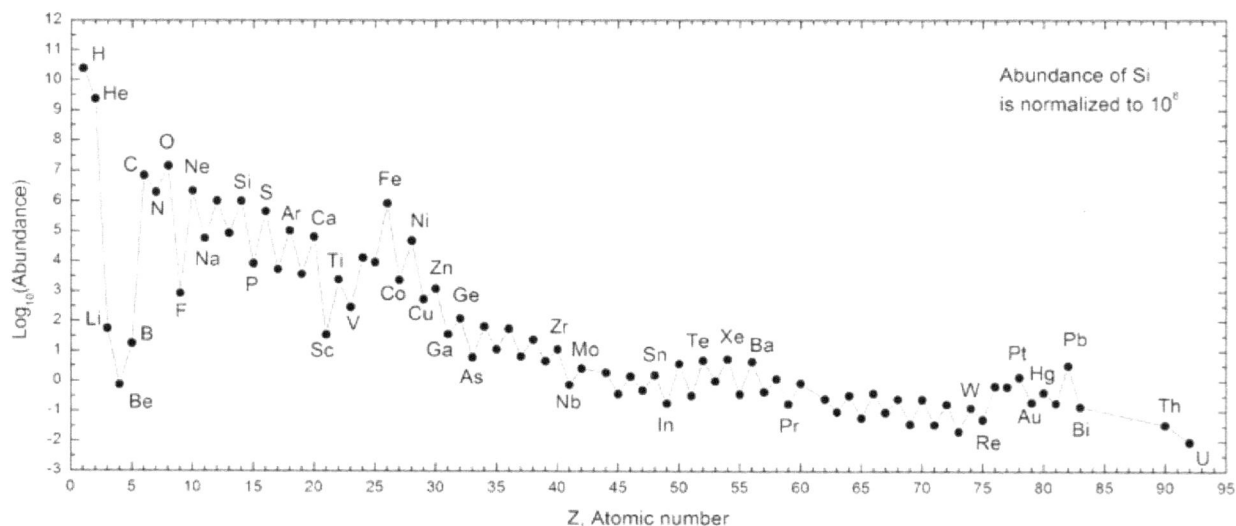

Abundances of the chemical elements in the Solar system. Hydrogen and helium are most common, from the Big Bang. The next three elements (Li, Be, B) are rare because they are poorly synthesized in the Big Bang and also in stars. The two general trends in the remaining stellar-produced elements are: (1) an alternation of abundance in elements as they have even or odd atomic numbers (the Oddo-Harkins rule), and (2) a general decrease in abundance as elements become heavier. Iron is especially common because it represents the minimum energy nuclide that can be made by fusion of helium in supernovae.

Elements in our galaxy	Parts per million by mass
Hydrogen	739,000
Helium	240,000
Oxygen	10,400
Carbon	4,600
Neon	1,340
Iron	1,090
Nitrogen	960
Silicon	650
Magnesium	580
Sulfur	440
Potassium	210
Nickel	100

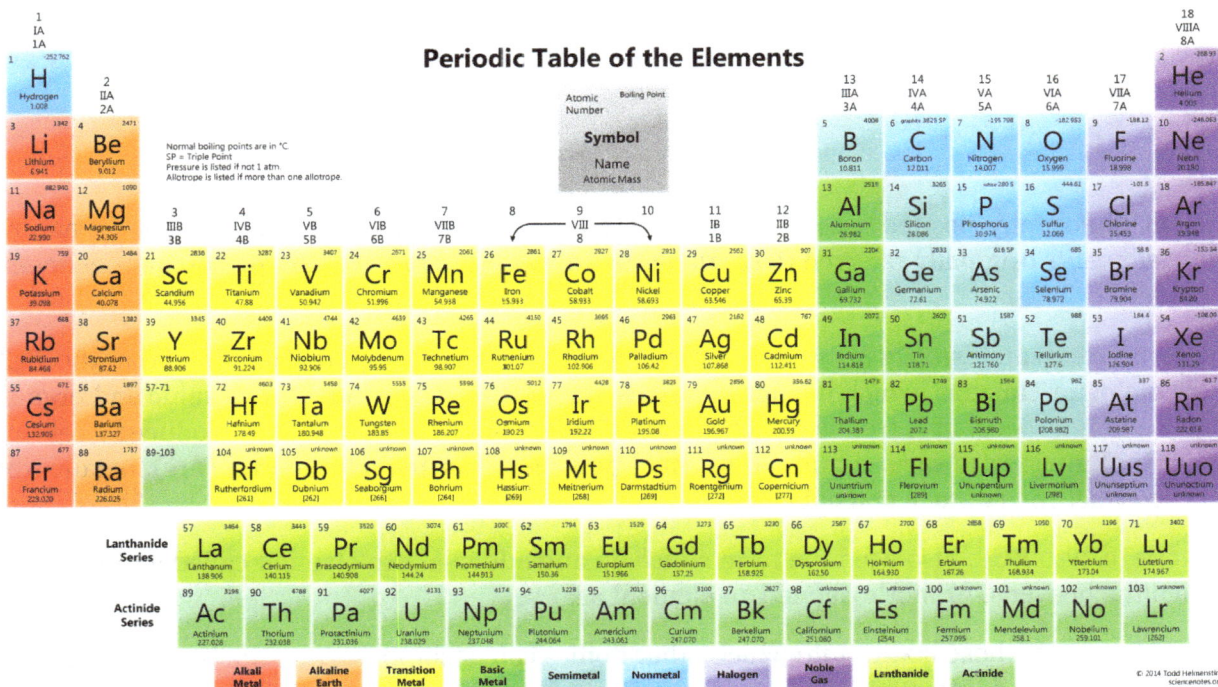

Periodic Table of the Elements

History

ОПЫТЪ СИСТЕМЫ ЭЛЕМЕНТОВЪ,

ОСНОВАННОЙ НА ИХЪ АТОМНОМЪ ВѢСѢ И ХИМИЧЕСКОМЪ СХОДСТВѢ.

			Ti=50	Zr=90	?=180.
			V=51	Nb=94	Ta=182.
			Cr=52	Mo=96	W=186.
			Mn=55	Rh=104,4	Pt=197,1.
			Fe=56	Ru=104,4	Ir=198.
		Ni=Co=59		Pd=106,6	Os=199.
H=1			Cu=63,4	Ag=108	Hg=200.
	Be= 9,4	Mg=24	Zn=65,2	Cd=112	
	B=11	Al=27,3	?=68	Ur=116	Au=197?
	C=12	Si=28	?=70	Sn=118	
	N=14	P=31	As=75	Sb=122	Bi=210?
	O=16	S=32	Se=79,4	Te=128?	
	F=19	Cl=35,5	Br=80	I=127	
Li=7	Na=23	K=39	Rb=85,4	Cs=133	Tl=204.
		Ca=40	Sr=87,6	Ba=137	Pb=207.
		?=45	Ce=92		
		?Er=56	La=94		
		?Yt=60	Di=95		
		?In=75,6	Th=118?		

Д. Менделѣевъ

Mendeleev's 1869 periodic table: *An experiment on a system of elements. Based on their atomic weights and chemical similarities.*

Evolving Definitions

The concept of an "element" as an undivisible substance has developed through three major historical phases: Classical definitions (such as those of the ancient Greeks), chemical definitions, and atomic definitions.

Classical Definitions

Ancient philosophy posited a set of classical elements to explain observed patterns in nature. These *elements* originally referred to *earth*, *water*, *air* and *fire* rather than the chemical elements of modern science.

The term 'elements' (*stoicheia*) was first used by the Greek philosopher Plato in about 360 BCE in his dialogue Timaeus, which includes a discussion of the composition of inorganic and organic bodies and is a speculative treatise on chemistry. Plato believed the elements introduced a century earlier by Empedocles were composed of small polyhedral forms: tetrahedron (fire), octahedron (air), icosahedron (water), and cube (earth).

Aristotle, c. 350 BCE, also used the term *stoicheia* and added a fifth element called aether, which formed the heavens. Aristotle defined an element as:

Element – one of those bodies into which other bodies can decompose, and that itself is not capable of being divided into other.

Chemical Definitions

In 1661, Robert Boyle proposed his theory of corpuscularism which favoured the analysis of matter as constituted by irreducible units of matter (atoms) and, choosing to side with neither Aristotle's view of the four elements nor Paracelsus' view of three fundamental elements, left open the question of the number of elements. The first modern list of chemical elements was given in Antoine Lavoisier's 1789 *Elements of Chemistry*, which contained thirty-three elements, including light and caloric. By 1818, Jöns Jakob Berzelius had determined atomic weights for forty-five of the forty-nine then-accepted elements. Dmitri Mendeleev had sixty-six elements in his periodic table of 1869.

Dmitri Mendeleev

From Boyle until the early 20th century, an element was defined as a pure substance that could not be decomposed into any simpler substance. Put another way, a chemical element cannot be transformed into other chemical elements by chemical processes. Elements during this time were generally distinguished by their atomic weights, a property measurable with fair accuracy by available analytical techniques.

Atomic Definitions

Henry Moseley

The 1913 discovery by English physicist Henry Moseley that the nuclear charge is the physical basis for an atom's atomic number, further refined when the nature of protons and neutrons became appreciated, eventually led to the current definition of an element based on atomic number (number of protons per atomic nucleus). The use of atomic numbers, rather than atomic weights, to distinguish elements has greater predictive value (since these numbers are integers), and also resolves some ambiguities in the chemistry-based view due to varying properties of isotopes and allotropes within the same element. Currently, IUPAC defines an element to exist if it has isotopes with a lifetime longer than the 10^{-14} seconds it takes the nucleus to form an electronic cloud.

By 1914, seventy-two elements were known, all naturally occurring. The remaining naturally occurring elements were discovered or isolated in subsequent decades, and various additional elements have also been produced synthetically, with much of that work pioneered by Glenn T. Seaborg. In 1955, element 101 was discovered and named mendelevium in honor of D.I. Mendeleev, the first to arrange the elements in a periodic manner. Most recently, the synthesis of element 118 was reported in October 2006, and the synthesis of element 117 was reported in April 2010.

Discovery and Recognition of Various Elements

Ten materials familiar to various prehistoric cultures are now known to be chemical elements: Carbon, copper, gold, iron, lead, mercury, silver, sulfur, tin, and zinc. Three additional materials now accepted as elements, arsenic, antimony, and bismuth, were recognized as distinct substances prior to 1500 AD. Phosphorus, cobalt, and platinum were isolated before 1750.

Most of the remaining naturally occurring chemical elements were identified and characterized by 1900, including:

- Such now-familiar industrial materials as aluminium, silicon, nickel, chromium, magnesium, and tungsten

- Reactive metals such as lithium, sodium, potassium, and calcium

- The halogens fluorine, chlorine, bromine, and iodine

- Gases such as hydrogen, oxygen, nitrogen, helium, argon, and neon

- Most of the rare-earth elements, including cerium, lanthanum, gadolinium, and neodymium.

- The more common radioactive elements, including uranium, thorium, radium, and radon

Elements isolated or produced since 1900 include:

- The three remaining undiscovered regularly occurring stable natural elements: hafnium, lutetium, and rhenium

- Plutonium, which was first produced synthetically in 1940 by Glenn T. Seaborg, but is now also known from a few long-persisting natural occurrences

- The three incidentally occurring natural elements (neptunium, promethium, and technetium), which were all first produced synthetically but later discovered in trace amounts in certain geological samples

- Three scarce decay products of uranium or thorium, (astatine, francium, and protactinium), and

- Various synthetic transuranic elements, beginning with americium and curium

Recently Discovered Elements

The first transuranium element (element with atomic number greater than 92) discovered was neptunium in 1940. Since 1999 claims for the discovery of new elements have been considered by the IUPAC/IUPAP Joint Working Party. As of January 2016, all 118 elements have been confirmed as discovered by IUPAC. The discovery of element 112 was acknowledged in 2009, and the name *copernicium* and the atomic symbol *Cn* were suggested for it. The name and symbol were officially endorsed by IUPAC on 19 February 2010. The heaviest element that is believed to have been synthesized to date is element 118, ununoctium, on 9 October 2006, by the Flerov Laboratory of Nuclear Reactions in Dubna, Russia. Element 117 was the latest element claimed to be discovered, in 2009. IUPAC officially recognized flerovium and livermorium, elements 114 and 116, in June 2011 and approved their names in May 2012. In December 2015, IUPAC recognized elements 113, 115, 117 and 118, and announced the elements' proposed final names on 8 June 2016. The names, nihonium (113, Nh), moscovium (115, Mc), tennessine (117, Ts), and oganesson (118, Og), are expected to be approved by the end of 2016.

List of the 118 Known Chemical Elements

The following sortable table includes the 118 known chemical elements.

- Atomic number, name, and symbol all serve independently as unique identifiers.

- Names are those accepted by IUPAC; provisional names for recently produced elements not yet formally named are in parentheses.

- Group, period, and block refer to an element's position in the periodic table. Group numbers here show the currently accepted numbering; for older alternate numberings, see Group (periodic table).

- State of matter (solid, liquid, or gas) applies at standard temperature and pressure conditions (STP).

- Occurrence distinguishes naturally occurring elements, categorized as either primordial or transient (from decay), and additional synthetic elements that have been produced technologically, but are not known to occur naturally.

- Description summarizes an element's properties using the broad categories commonly presented in periodic tables: Actinide, alkali metal, alkaline earth metal, lanthanide, post-transition metal, metalloid, noble gas, polyatomic or diatomic nonmetal, and transition metal.

						State at STP	Occur-rence	
Atomic no.	Name	Sym-bol	Group	Period	Block			Description
1	Hydrogen	H	1	1	s	Gas	Primordial	Diatomic nonmetal
2	Helium	He	18	1	s	Gas	Primordial	Noble gas
3	Lithium	Li	1	2	s	Solid	Primordial	Alkali metal
4	Beryllium	Be	2	2	s	Solid	Primordial	Alkaline earth metal
5	Boron	B	13	2	p	Solid	Primordial	Metalloid
6	Carbon	C	14	2	p	Solid	Primordial	Polyatomic nonmetal
7	Nitrogen	N	15	2	p	Gas	Primordial	Diatomic nonmetal
8	Oxygen	O	16	2	p	Gas	Primordial	Diatomic nonmetal
9	Fluorine	F	17	2	p	Gas	Primordial	Diatomic nonmetal
10	Neon	Ne	18	2	p	Gas	Primordial	Noble gas
11	Sodium	Na	1	3	s	Solid	Primordial	Alkali metal
12	Magnesium	Mg	2	3	s	Solid	Primordial	Alkaline earth metal
13	Aluminium	Al	13	3	p	Solid	Primordial	Post-transition metal
14	Silicon	Si	14	3	p	Solid	Primordial	Metalloid
15	Phosphorus	P	15	3	p	Solid	Primordial	Polyatomic nonmetal
16	Sulfur	S	16	3	p	Solid	Primordial	Polyatomic nonmetal
17	Chlorine	Cl	17	3	p	Gas	Primordial	Diatomic nonmetal
18	Argon	Ar	18	3	p	Gas	Primordial	Noble gas
19	Potassium	K	1	4	s	Solid	Primordial	Alkali metal
20	Calcium	Ca	2	4	s	Solid	Primordial	Alkaline earth metal
21	Scandium	Sc	3	4	d	Solid	Primordial	Transition metal
22	Titanium	Ti	4	4	d	Solid	Primordial	Transition metal
23	Vanadium	V	5	4	d	Solid	Primordial	Transition metal
24	Chromium	Cr	6	4	d	Solid	Primordial	Transition metal
25	Manganese	Mn	7	4	d	Solid	Primordial	Transition metal
26	Iron	Fe	8	4	d	Solid	Primordial	Transition metal
27	Cobalt	Co	9	4	d	Solid	Primordial	Transition metal
28	Nickel	Ni	10	4	d	Solid	Primordial	Transition metal
29	Copper	Cu	11	4	d	Solid	Primordial	Transition metal
30	Zinc	Zn	12	4	d	Solid	Primordial	Transition metal
31	Gallium	Ga	13	4	p	Solid	Primordial	Post-transition metal
32	Germanium	Ge	14	4	p	Solid	Primordial	Metalloid
33	Arsenic	As	15	4	p	Solid	Primordial	Metalloid
34	Selenium	Se	16	4	p	Solid	Primordial	Polyatomic nonmetal
35	Bromine	Br	17	4	p	Liquid	Primordial	Diatomic nonmetal
36	Krypton	Kr	18	4	p	Gas	Primordial	Noble gas
37	Rubidium	Rb	1	5	s	Solid	Primordial	Alkali metal
38	Strontium	Sr	2	5	s	Solid	Primordial	Alkaline earth metal
39	Yttrium	Y	3	5	d	Solid	Primordial	Transition metal

List of elements

40	Zirconium	Zr	4	5	d	Solid	Primordial	Transition metal
41	Niobium	Nb	5	5	d	Solid	Primordial	Transition metal
42	Molybdenum	Mo	6	5	d	Solid	Primordial	Transition metal
43	Technetium	Tc	7	5	d	Solid	Transient	Transition metal
44	Ruthenium	Ru	8	5	d	Solid	Primordial	Transition metal
45	Rhodium	Rh	9	5	d	Solid	Primordial	Transition metal
46	Palladium	Pd	10	5	d	Solid	Primordial	Transition metal
47	Silver	Ag	11	5	d	Solid	Primordial	Transition metal
48	Cadmium	Cd	12	5	d	Solid	Primordial	Transition metal
49	Indium	In	13	5	p	Solid	Primordial	Post-transition metal
50	Tin	Sn	14	5	p	Solid	Primordial	Post-transition metal
51	Antimony	Sb	15	5	p	Solid	Primordial	Metalloid
52	Tellurium	Te	16	5	p	Solid	Primordial	Metalloid
53	Iodine	I	17	5	p	Solid	Primordial	Diatomic nonmetal
54	Xenon	Xe	18	5	p	Gas	Primordial	Noble gas
55	Caesium	Cs	1	6	s	Solid	Primordial	Alkali metal
56	Barium	Ba	2	6	s	Solid	Primordial	Alkaline earth metal
57	Lanthanum	La	3	6	f	Solid	Primordial	Lanthanide
58	Cerium	Ce	3	6	f	Solid	Primordial	Lanthanide
59	Praseodymium	Pr	3	6	f	Solid	Primordial	Lanthanide
60	Neodymium	Nd	3	6	f	Solid	Primordial	Lanthanide
61	Promethium	Pm	3	6	f	Solid	Transient	Lanthanide
62	Samarium	Sm	3	6	f	Solid	Primordial	Lanthanide
63	Europium	Eu	3	6	f	Solid	Primordial	Lanthanide
64	Gadolinium	Gd	3	6	f	Solid	Primordial	Lanthanide
65	Terbium	Tb	3	6	f	Solid	Primordial	Lanthanide
66	Dysprosium	Dy	3	6	f	Solid	Primordial	Lanthanide
67	Holmium	Ho	3	6	f	Solid	Primordial	Lanthanide
68	Erbium	Er	3	6	f	Solid	Primordial	Lanthanide
69	Thulium	Tm	3	6	f	Solid	Primordial	Lanthanide
70	Ytterbium	Yb	3	6	f	Solid	Primordial	Lanthanide
71	Lutetium	Lu	3	6	d	Solid	Primordial	Lanthanide
72	Hafnium	Hf	4	6	d	Solid	Primordial	Transition metal
73	Tantalum	Ta	5	6	d	Solid	Primordial	Transition metal
74	Tungsten	W	6	6	d	Solid	Primordial	Transition metal
75	Rhenium	Re	7	6	d	Solid	Primordial	Transition metal
76	Osmium	Os	8	6	d	Solid	Primordial	Transition metal
77	Iridium	Ir	9	6	d	Solid	Primordial	Transition metal
78	Platinum	Pt	10	6	d	Solid	Primordial	Transition metal
79	Gold	Au	11	6	d	Solid	Primordial	Transition metal
80	Mercury	Hg	12	6	d	Liquid	Primordial	Transition metal
81	Thallium	Tl	13	6	p	Solid	Primordial	Post-transition metal

82	Lead	Pb	14	6	p	Solid	Primordial	Post-transition metal
83	Bismuth	Bi	15	6	p	Solid	Primordial	Post-transition metal
84	Polonium	Po	16	6	p	Solid	Transient	Post-transition metal
85	Astatine	At	17	6	p	Solid	Transient	Metalloid
86	Radon	Rn	18	6	p	Gas	Transient	Noble gas
87	Francium	Fr	1	7	s	Solid	Transient	Alkali metal
88	Radium	Ra	2	7	s	Solid	Transient	Alkaline earth metal
89	Actinium	Ac	3	7	f	Solid	Transient	Actinide
90	Thorium	Th	3	7	f	Solid	Primordial	Actinide
91	Protactinium	Pa	3	7	f	Solid	Transient	Actinide
92	Uranium	U	3	7	f	Solid	Primordial	Actinide
93	Neptunium	Np	3	7	f	Solid	Transient	Actinide
94	Plutonium	Pu	3	7	f	Solid	Primordial	Actinide
95	Americium	Am	3	7	f	Solid	Synthetic	Actinide
96	Curium	Cm	3	7	f	Solid	Synthetic	Actinide
97	Berkelium	Bk	3	7	f	Solid	Synthetic	Actinide
98	Californium	Cf	3	7	f	Solid	Synthetic	Actinide
99	Einsteinium	Es	3	7	f	Solid	Synthetic	Actinide
100	Fermium	Fm	3	7	f		Synthetic	Actinide
101	Mendelevium	Md	3	7	f		Synthetic	Actinide
102	Nobelium	No	3	7	f		Synthetic	Actinide
103	Lawrencium	Lr	3	7	d		Synthetic	Actinide
104	Rutherfordium	Rf	4	7	d		Synthetic	Transition metal
105	Dubnium	Db	5	7	d		Synthetic	Transition metal
106	Seaborgium	Sg	6	7	d		Synthetic	Transition metal
107	Bohrium	Bh	7	7	d		Synthetic	Transition metal
108	Hassium	Hs	8	7	d		Synthetic	Transition metal
109	Meitnerium	Mt	9	7	d		Synthetic	
110	Darmstadtium	Ds	10	7	d		Synthetic	
111	Roentgenium	Rg	11	7	d		Synthetic	
112	Copernicium	Cn	12	7	d		Synthetic	Transition metal
113	(Ununtrium)	(Uut)	13	7	p		Synthetic	
114	Flerovium	Fl	14	7	p		Synthetic	Post-transition metal
115	(Ununpentium)	(Uup)	15	7	p		Synthetic	
116	Livermorium	Lv	16	7	p		Synthetic	
117	(Ununseptium)	(Uus)	17	7	p		Synthetic	
118	(Ununoctium)	(Uuo)	18	7	p		Synthetic	

References

- R. Penrose (1991). "The mass of the classical vacuum". In S. Saunders, H.R. Brown. The Philosophy of Vacuum. Oxford University Press. p. 21. ISBN 0-19-824449-5.

- J. Olmsted; G.M. Williams (1996). Chemistry: The Molecular Science (2nd ed.). Jones & Bartlett. p. 40. ISBN 0-8151-8450-6

- K.A. Peacock (2008). The Quantum Revolution: A Historical Perspective. Greenwood Publishing Group. p. 47. ISBN 0-313-33448-X.

- M.H. Krieger (1998). Constitutions of Matter: Mathematically Modeling the Most Everyday of Physical Phenomena. University of Chicago Press. p. 22. ISBN 0-226-45305-7.

- G. F. Barker (1870). "Divisions of matter". A text-book of elementary chemistry: theoretical and inorganic. John F Morton & Co. p. 2. ISBN 978-1-4460-2206-1.

- B. Povh; K. Rith; C. Scholz; F. Zetsche; M. Lavelle (2004). "Part I: Analysis: The building blocks of matter". Particles and Nuclei: An Introduction to the Physical Concepts (4th ed.). Springer. ISBN 3-540-20168-8.

- L. Smolin (2007). The Trouble with Physics: The Rise of String Theory, the Fall of a Science, and What Comes Next. Mariner Books. p. 67. ISBN 0-618-91868-X.

- B. Povh; K. Rith; C. Scholz; F. Zetsche; M. Lavelle (2004). Particles and Nuclei: An Introduction to the Physical Concepts. Springer. p. 103. ISBN 3-540-20168-8.

- T. Hatsuda (2008). "Quark–gluon plasma and QCD". In H. Akai. Condensed matter theories 21. Nova Publishers. p. 296. ISBN 1-60021-501-7.

- K.W Staley (2004). "Origins of the Third Generation of Matter". The Evidence for the Top Quark. Cambridge University Press. p. 8. ISBN 0-521-82710-8.

- H.S. Goldberg; M.D. Scadron (1987). Physics of Stellar Evolution and Cosmology. Taylor & Francis. p. 202. ISBN 0-677-05540-4.

- P.J. Collings (2002). "Chapter 1: States of Matter". Liquid Crystals: Nature's Delicate Phase of Matter. Princeton University Press. ISBN 0-691-08672-9.

- National Research Council (US) (2006). Revealing the hidden nature of space and time. National Academies Press. p. 46. ISBN 0-309-10194-8.

- K. Pretzl (2004). "Dark Matter, Massive Neutrinos and Susy Particles". Structure and Dynamics of Elementary Matter. Walter Greiner. p. 289. ISBN 1-4020-2446-0.

- K. Freeman; G. McNamara (2006). "What can the matter be?". In Search of Dark Matter. Birkhäuser Verlag. p. 105. ISBN 0-387-27616-5.

- J.C. Wheeler (2007). Cosmic Catastrophes: Exploding Stars, Black Holes, and Mapping the Universe. Cambridge University Press. p. 282. ISBN 0-521-85714-7.

- J. Gribbin (2007). The Origins of the Future: Ten Questions for the Next Ten Years. Yale University Press. p. 151. ISBN 0-300-12596-8.

- M. H. Jones; R. J. Lambourne; D. J. Adams (2004). An Introduction to Galaxies and Cosmology. Cambridge University Press. p. 21; Fig. 1.13. ISBN 0-521-54623-0.

- M. Wenham (2005). Understanding Primary Science: Ideas, Concepts and Explanations (2nd ed.). Paul Chapman Educational Publishing. p. 115. ISBN 1-4129-0163-4.

- T.H. Levere (1993). "Introduction". Affinity and Matter: Elements of Chemical Philosophy, 1800–1865. Taylor & Francis. ISBN 2-88124-583-8.

- B.A. Schumm (2004). Deep Down Things: The Breathtaking Beauty of Particle Physics. Johns Hopkins University Press. p. 57. ISBN 0-8018-7971-X.

- P. Schmüser; H. Spitzer (2002). "Particles". In L. Bergmann; et al. Constituents of Matter: Atoms, Molecules, Nuclei. CRC Press. pp. 773 ff. ISBN 0-8493-1202-7.

- P. M. Chaikin; T. C. Lubensky (2000). Principles of Condensed Matter Physics. Cambridge University Press. p. xvii. ISBN 0-521-79450-1.

Subdisciplines of Chemistry

Chemistry is an interdisciplinary subject. This content will provide a glimpse of related fields of chemistry briefly. Biochemistry, physical chemistry and organic chemistry are some of the various branches of chemistry that are explained in this chapter. It provides a plethora of interdisciplinary topics for better comprehension of chemistry.

Biochemistry

Biochemistry, sometimes called biological chemistry, is the study of chemical processes within and relating to living organisms. By controlling information flow through biochemical signaling and the flow of chemical energy through metabolism, biochemical processes give rise to the complexity of life. Over the last decades of the 20th century, biochemistry has become so successful at explaining living processes that now almost all areas of the life sciences from botany to medicine to genetics are engaged in biochemical research. Today, the main focus of pure biochemistry is on understanding how biological molecules give rise to the processes that occur within living cells, which in turn relates greatly to the study and understanding of tissues, organs, and whole organisms—that is, all of biology.

Biochemistry is closely related to molecular biology, the study of the molecular mechanisms by which genetic information encoded in DNA is able to result in the processes of life. Depending on the exact definition of the terms used, molecular biology can be thought of as a branch of biochemistry, or biochemistry as a tool with which to investigate and study molecular biology.

Much of biochemistry deals with the structures, functions and interactions of biological macromolecules, such as proteins, nucleic acids, carbohydrates and lipids, which provide the structure of cells and perform many of the functions associated with life. The chemistry of the cell also depends on the reactions of smaller molecules and ions. These can be inorganic, for example water and metal ions, or organic, for example the amino acids, which are used to synthesize proteins. The mechanisms by which cells harness energy from their environment via chemical reactions are known as metabolism. The findings of biochemistry are applied primarily in medicine, nutrition, and agriculture. In medicine, biochemists investigate the causes and cures of diseases. In nutrition, they study how to maintain health and study the effects of nutritional deficiencies. In agriculture, biochemists investigate soil and fertilizers, and try to discover ways to improve crop cultivation, crop storage and pest control.

History

At its broadest definition, biochemistry can be seen as a study of the components and composition of living things and how they come together to become life, and the history of biochemistry may therefore go back as far as the ancient Greeks. However, biochemistry as a specific scientific discipline has its beginning some time in the 19th century, or a little earlier, depending on which aspect of biochemistry one is being focused on. Some argued that the beginning of biochemistry may have been the discovery of the first enzyme, diastase (today called amylase), in 1833 by Anselme Payen, while others considered Eduard Buchner's first demonstration of a complex biochemical process alcoholic fermentation in cell-free extracts in 1897 to be the birth of biochemistry. Some might also point as its beginning to the influential 1842 work by Justus von Liebig, *Animal chemistry, or, Organic chemistry in its applications to physiology and pathology*, which presented a chemical theory of metabolism, or even earlier to the 18th century studies on fermentation and respiration by Antoine Lavoisier. Many other pioneers in the field who helped to uncover the layers of complexity of biochemistry have been proclaimed founders of modern biochemistry, for example Emil Fischer for his work on the chemistry of proteins, and F. Gowland Hopkins on enzymes and the dynamic nature of biochemistry.

Gerty Cori and Carl Cori jointly won the Nobel Prize in 1947 for their discovery of the Cori cycle at RPMI.

The term "biochemistry" itself is derived from a combination of biology and chemistry. In 1877, Felix Hoppe-Seyler used the term (*biochemie* in German) as a synonym for physiological chemistry in the foreword to the first issue of *Zeitschrift für Physiologische Chemie* (Journal of Physiological Chemistry) where he argued for the setting up of institutes dedicated to this field of study. The German chemist Carl Neuberg however is often cited to have been coined the word in 1903, while some credited it to Franz Hofmeister.

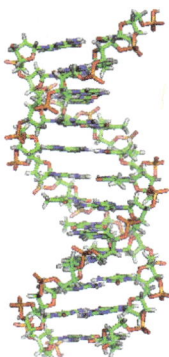

DNA structure (1D65)

It was once generally believed that life and its materials had some essential property or substance (often referred to as the "vital principle") distinct from any found in non-living matter, and it was thought that only living beings could produce the molecules of life. Then, in 1828, Friedrich Wöhler published a paper on the synthesis of urea, proving that organic compounds can be created artificially. Since then, biochemistry has advanced, especially since the mid-20th century, with the development of new techniques such as chromatography, X-ray diffraction, dual polarisation interferometry, NMR spectroscopy, radioisotopic labeling, electron microscopy, and molecular dynamics simulations. These techniques allowed for the discovery and detailed analysis of many molecules and metabolic pathways of the cell, such as glycolysis and the Krebs cycle (citric acid cycle).

Another significant historic event in biochemistry is the discovery of the gene and its role in the transfer of information in the cell. This part of biochemistry is often called molecular biology. In the 1950s, James D. Watson, Francis Crick, Rosalind Franklin, and Maurice Wilkins were instrumental in solving DNA structure and suggesting its relationship with genetic transfer of information. In 1958, George Beadle and Edward Tatum received the Nobel Prize for work in fungi showing that one gene produces one enzyme. In 1988, Colin Pitchfork was the first person convicted of murder with DNA evidence, which led to growth of forensic science. More recently, Andrew Z. Fire and Craig C. Mello received the 2006 Nobel Prize for discovering the role of RNA interference (RNAi), in the silencing of gene expression.

Starting Materials: The Chemical Elements of Life

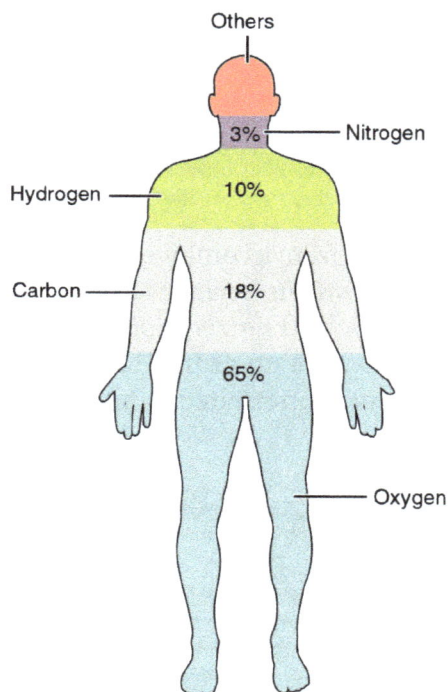

The main elements that compose the human body are shown from most abundant (by mass) to least abundant.

Around two dozen of the 92 naturally occurring chemical elements are essential to various kinds of biological life. Most rare elements on Earth are not needed by life (exceptions being selenium and iodine), while a few common ones (aluminum and titanium) are not used. Most organisms share element needs, but there are a few differences between plants and animals. For example, ocean algae use bromine, but land plants and animals seem to need none. All animals require sodium, but some plants do not. Plants need boron and silicon, but animals may not (or may need ultra-small amounts).

Just six elements—carbon, hydrogen, nitrogen, oxygen, calcium, and phosphorus—make up almost 99% of the mass of living cells, including those in the human body. In addition to the six major elements that compose most of the human body, humans require smaller amounts of possibly 18 more.

Biomolecules

The four main classes of molecules in biochemistry (often called biomolecules) are carbohydrates, lipids, proteins, and nucleic acids. Many biological molecules are polymers: in this terminology, *monomers* are relatively small micromolecules that are linked together to create large macro-molecules known as *polymers*. When monomers are linked together to synthesize a biological polymer, they undergo a process called dehydration synthesis. Different macro-molecules can as-semble in larger complexes, often needed for biological activity.

Carbohydrates

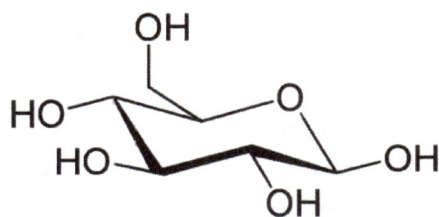

Glucose, a monosaccharide

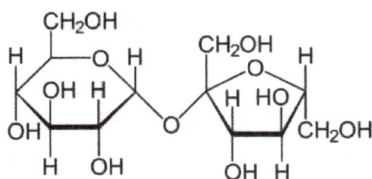

A molecule of sucrose (glucose + fructose), a disaccharide

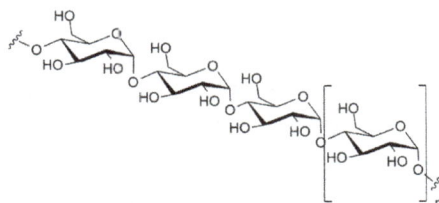

Amylose, a polysaccharide made up of several thousand glucose units

The function of carbohydrates includes energy storage and providing structure. Sugars are carbohydrates, but not all carbohydrates are sugars. There are more carbohydrates on Earth than any other known type of biomolecule; they are used to store energy and genetic information, as well as play important roles in cell to cell interactions and communications.

The simplest type of carbohydrate is a monosaccharide, which among other properties contains carbon, hydrogen, and oxygen, mostly in a ratio of 1:2:1 (generalized formula $C_nH_{2n}O_n$, where n is at least 3). Glucose ($C_6H_{12}O_6$) is one of the most important carbohydrates, others include fructose ($C_6H_{12}O_6$), the sugar commonly associated with the sweet taste of fruits,[a] and deoxyribose ($C_5H_{10}O_4$).

A monosaccharide can switch from the acyclic (open-chain) form to a cyclic form, through a nucleophilic addition reaction between the carbonyl group and one of the hydroxyls of the same molecule. The reaction creates a ring of carbon atoms closed by one bridging oxygen atom. The resulting molecule has an hemiacetal or hemiketal group, depending on whether the linear form was an aldose or a ketose. The reaction is easily reversed, yielding the original open-chain form.

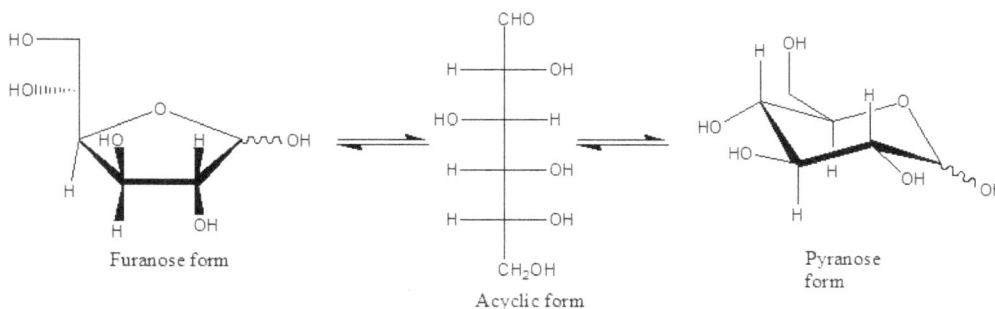

Conversion between the furanose, acyclic, and pyranose forms of D-glucose.

In these cyclic forms, the ring usually has **5** or **6** atoms. These forms are called furanoses and pyranoses, respectively — by analogy with furan and pyran, the simplest compounds with the same carbon-oxygen ring (although they lack the double bonds of these two molecules). For example, the aldohexose glucose may form a hemiacetal linkage between the hydroxyl on carbon 1 and the oxygen on carbon 4, yielding a molecule with a 5-membered ring, called glucofuranose. The same reaction can take place between carbons 1 and 5 to form a molecule with a 6-membered ring, called glucopyranose. Cyclic forms with a 7-atom ring (the same of oxepane), rarely encountered, are called heptoses.

When two monosaccharides undergo dehydration synthesis whereby a molecule of water is released, as two hydrogen atoms and one oxygen atom are lost from the two monosaccharides. The new molecule, consisting of two monosaccharides, is called a *disaccharide* and is conjoined together by a glycosidic or ether bond. The reverse reaction can also occur, using a molecule of water to split up a disaccharide and break the glycosidic bond; this is termed *hydrolysis*. The most well-known disaccharide is sucrose, ordinary sugar (in scientific contexts, called *table sugar* or *cane sugar* to differentiate it from other sugars). Sucrose consists of a glucose molecule and a fructose molecule joined together. Another important disaccharide is lactose, consisting of a glucose molecule and a galactose molecule. As most humans age, the production of lactase, the enzyme that hydrolyzes lactose back into glucose and galactose, typically decreases. This results in lactase deficiency, also called *lactose intolerance.*

When a few (around three to six) monosaccharides are joined, it is called an *oligosaccharide* (*oligo-* meaning "few"). These molecules tend to be used as markers and signals, as well as having some other uses. Many monosaccharides joined together make a polysaccharide. They can be joined together in one long linear chain, or they may be branched. Two of the most common polysaccharides are cellulose and glycogen, both consisting of repeating glucose monomers. Examples are *Cellulose* which is an important structural component of plant's cell walls, and *glycogen*, used as a form of energy storage in animals.

Sugar can be characterized by having reducing or non-reducing ends. A reducing end of a carbohydrate is a carbon atom that can be in equilibrium with the open-chain aldehyde (aldose) or keto form (ketose). If the joining of monomers takes place at such a carbon atom, the free hydroxy group of the pyranose or furanose form is exchanged with an OH-side-chain of another sugar, yielding a full acetal. This prevents opening of the chain to the aldehyde or keto form and renders the modified residue non-reducing. Lactose contains a reducing end at its glucose moiety, whereas the galactose moiety form a full acetal with the C4-OH group of glucose. Saccharose does not have a reducing end because of full acetal formation between the aldehyde carbon of glucose (C1) and the keto carbon of fructose (C2).

Lipids

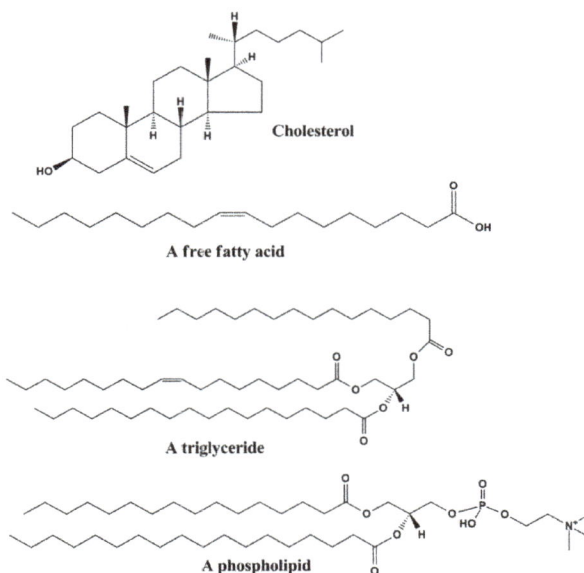

Structures of some common lipids. At the top are cholesterol and oleic acid. The middle structure is a triglyceride composed of oleoyl, stearoyl, and palmitoyl chains attached to a glycerol backbone. At the bottom is the common phospholipid, phosphatidylcholine.

Lipids comprises a diverse range of molecules and to some extent is a catchall for relatively water-insoluble or nonpolar compounds of biological origin, including waxes, fatty acids, fatty-acid derived phospholipids, sphingolipids, glycolipids, and terpenoids (e.g., retinoids and steroids). Some lipids are linear aliphatic molecules, while others have ring structures. Some are aromatic, while others are not. Some are flexible, while others are rigid.

Lipids are usually made from one molecule of glycerol combined with other molecules. In triglycerides, the main group of bulk lipids, there is one molecule of glycerol and three fatty acids. Fatty acids are considered the monomer in that case, and may be saturated (no double bonds in the carbon chain) or unsaturated (one or more double bonds in the carbon chain).

Most lipids have some polar character in addition to being largely nonpolar. In general, the bulk of their structure is nonpolar or hydrophobic ("water-fearing"), meaning that it does not interact well with polar solvents like water. Another part of their structure is polar or hydrophilic ("water-loving") and will tend to associate with polar solvents like water. This makes them amphiphilic molecules (having both hydrophobic and hydrophilic portions). In the case of cholesterol, the polar group is a mere -OH (hydroxyl or alcohol). In the case of phospholipids, the polar groups are considerably larger and more polar, as described below.

Lipids are an integral part of our daily diet. Most oils and milk products that we use for cooking and eating like butter, cheese, ghee etc., are composed of fats. Vegetable oils are rich in various polyunsaturated fatty acids (PUFA). Lipid-containing foods undergo digestion within the body and are broken into fatty acids and glycerol, which are the final degradation products of fats and lipids. Lipids, especially phospholipids, are also used in various pharmaceutical products, either as co-solubilisers (e.g., in parenteral infusions) or else as drug carrier components (e.g., in a liposome or transfersome).

Proteins

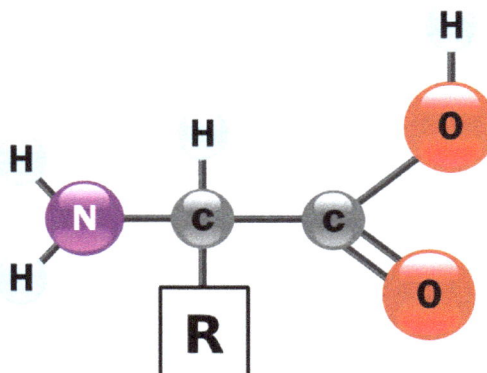

The general structure of an α-amino acid, with the amino group on the left and the carboxyl group on the right.

Proteins are very large molecules – macro-biopolymers – made from monomers called amino acids. An amino acid consists of a carbon atom bound to four groups. One is an amino group, $-NH_2$, and one is a carboxylic acid group, $-COOH$ (although these exist as $-NH_3^+$ and $-COO^-$ under physiologic conditions). The third is a simple hydrogen atom. The fourth is commonly denoted "$-R$" and is different for each amino acid. There are 20 standard amino acids, each containing a carboxyl group, an amino group, and a side-chain (known as an "R" group). The "R" group is what makes each amino acid different, and the properties of the side-chains greatly influence the overall three-dimensional conformation of a protein. Some amino acids have functions by themselves or in a modified form; for instance, glutamate functions as an important neurotransmitter. Amino acids can be joined via a peptide bond. In this dehydration synthesis, a water molecule is removed and the peptide bond connects the nitrogen of one amino acid's amino group to the carbon of the other's carboxylic acid group. The resulting molecule is called a *dipeptide*, and short stretches of amino acids (usually, fewer than thirty) are called *peptides* or polypeptides. Longer stretches merit the title *proteins*. As an example, the important blood serum protein albumin contains 585 amino acid residues.

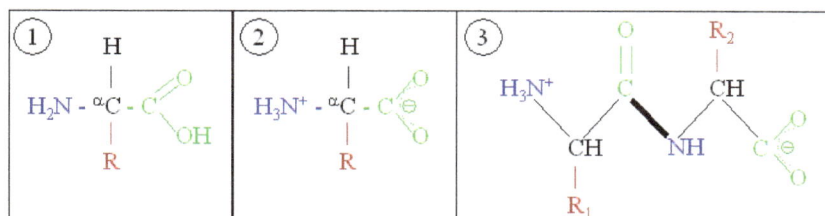

Generic amino acids (1) in neutral form, (2) as they exist physiologically, and (3) joined together as a dipeptide.

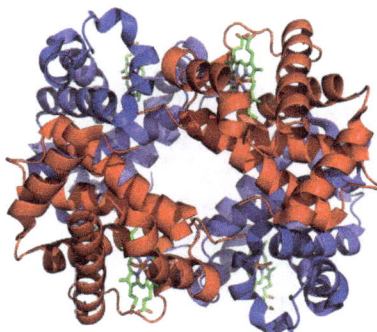

A schematic of hemoglobin. The red and blue ribbons represent the protein globin; the green structures are the heme groups.

Some proteins perform largely structural roles. For instance, movements of the proteins actin and myosin ultimately are responsible for the contraction of skeletal muscle. One property many proteins have is that they specifically bind to a certain molecule or class of molecules—they may be *extremely* selective in what they bind. Antibodies are an example of proteins that attach to one specific type of molecule. In fact, the enzyme-linked immunosorbent assay (ELISA), which uses antibodies, is one of the most sensitive tests modern medicine uses to detect various biomolecules. Probably the most important proteins, however, are the enzymes. Virtually every reaction in a living cell requires an enzyme to lower the activation energy of the reaction. These molecules recognize specific reactant molecules called *substrates*; they then catalyze the reaction between them. By lowering the activation energy, the enzyme speeds up that reaction by a rate of 10^{11} or more; a reaction that would normally take over 3,000 years to complete spontaneously might take less than a second with an enzyme. The enzyme itself is not used up in the process, and is free to catalyze the same reaction with a new set of substrates. Using various modifiers, the activity of the enzyme can be regulated, enabling control of the biochemistry of the cell as a whole.

The structure of proteins is traditionally described in a hierarchy of four levels. The primary structure of a protein simply consists of its linear sequence of amino acids; for instance, "alanine-glycine-tryptophan-serine-glutamate-asparagine-glycine-lysine-...". Secondary structure is concerned with local morphology (morphology being the study of structure). Some combinations of amino acids will tend to curl up in a coil called an α-helix or into a sheet called a β-sheet; some α-helixes can be seen in the hemoglobin schematic above. Tertiary structure is the entire three-dimensional shape of the protein. This shape is determined by the sequence of amino acids. In fact, a single change can change the entire structure. The alpha chain of hemoglobin contains 146 amino acid residues; substitution of the glutamate residue at position 6 with a valine residue changes the behavior of hemoglobin so much that it results in sickle-cell disease. Finally, quaternary structure is concerned with the structure of a protein with multiple peptide subunits, like hemoglobin with its four subunits. Not all proteins have more than one subunit.

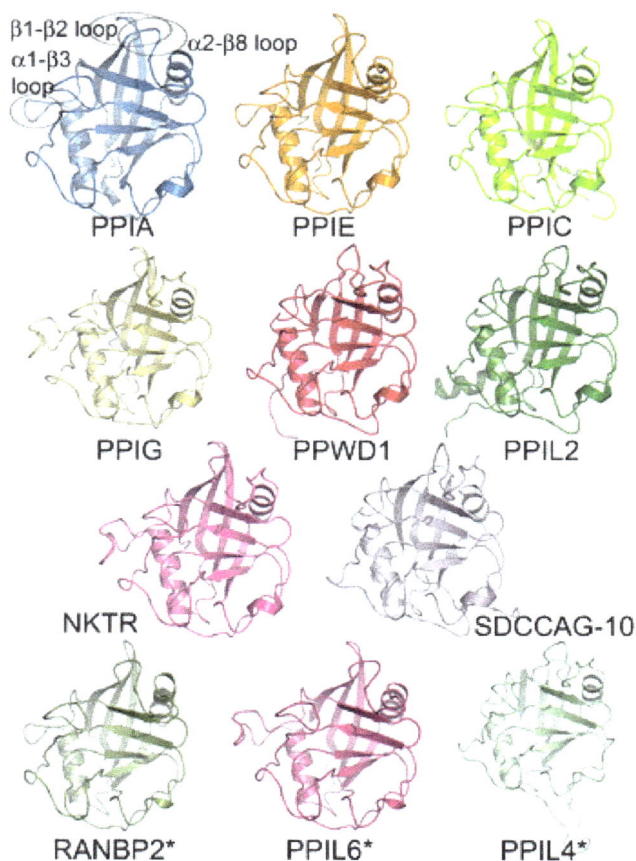

Members of a protein family, as represented by the structures of the isomerase domains.

Ingested proteins are usually broken up into single amino acids or dipeptides in the small intestine, and then absorbed. They can then be joined to make new proteins. Intermediate products of glycolysis, the citric acid cycle, and the pentose phosphate pathway can be used to make all twenty amino acids, and most bacteria and plants possess all the necessary enzymes to synthesize them. Humans and other mammals, however, can synthesize only half of them. They cannot synthesize isoleucine, leucine, lysine, methionine, phenylalanine, threonine, tryptophan, and valine. These are the essential amino acids, since it is essential to ingest them. Mammals do possess the enzymes to synthesize alanine, asparagine, aspartate, cysteine, glutamate, glutamine, glycine, proline, serine, and tyrosine, the nonessential amino acids. While they can synthesize arginine and histidine, they cannot produce it in sufficient amounts for young, growing animals, and so these are often considered essential amino acids.

If the amino group is removed from an amino acid, it leaves behind a carbon skeleton called an α-keto acid. Enzymes called transaminases can easily transfer the amino group from one amino acid (making it an α-keto acid) to another α-keto acid (making it an amino acid). This is important in the biosynthesis of amino acids, as for many of the pathways, intermediates from other biochemical pathways are converted to the α-keto acid skeleton, and then an amino group is added, often via transamination. The amino acids may then be linked together to make a protein.

A similar process is used to break down proteins. It is first hydrolyzed into its component amino acids. Free ammonia (NH_3), existing as the ammonium ion (NH_4^+) in blood, is toxic to life forms. A suitable method for excreting it must therefore exist. Different tactics have evolved in different animals, depending on the animals' needs. Unicellular organisms, of course, simply release the ammonia into the environment. Likewise, bony fish can release the ammonia into the water where it is quickly diluted. In general, mammals convert the ammonia into urea, via the urea cycle.

In order to determine whether two proteins are related, or in other words to decide whether they are homologous or not, scientists use sequence-comparison methods. Methods like sequence alignments and structural alignments are powerful tools that help scientists identify homologies between related molecules. The relevance of finding homologies among proteins goes beyond forming an evolutionary pattern of protein families. By finding how similar two protein sequences are, we acquire knowledge about their structure and therefore their function.

Nucleic Acids

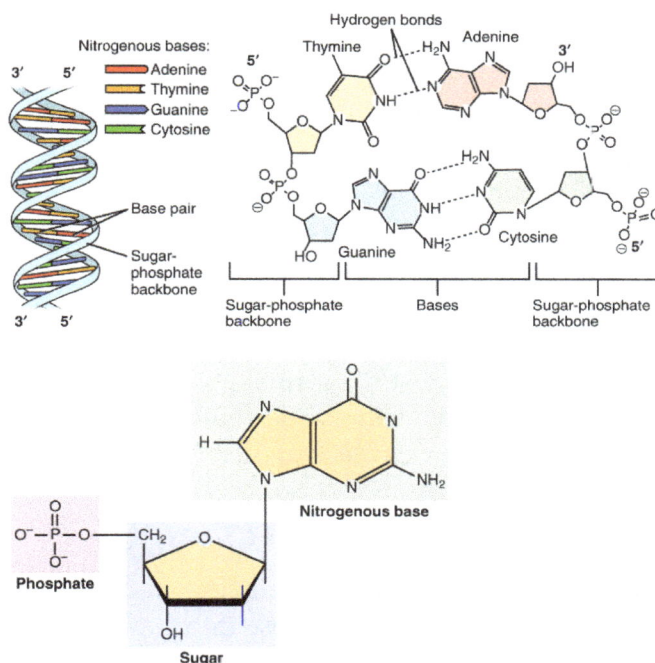

The structure of deoxyribonucleic acid (DNA), the picture shows the monomers being put together.

Nucleic acids, so called because of its prevalence in cellular nuclei, is the generic name of the family of biopolymers. They are complex, high-molecular-weight biochemical macromolecules that can convey genetic information in all living cells and viruses. The monomers are called nucleotides, and each consists of three components: a nitrogenous heterocyclic base (either a purine or a pyrimidine), a pentose sugar, and a phosphate group.

The most common nucleic acids are deoxyribonucleic acid (DNA) and ribonucleic acid (RNA). The phosphate group and the sugar of each nucleotide bond with each other to form the backbone of the nucleic acid, while the sequence of nitrogenous bases stores the information. The most common nitrogenous bases are adenine, cytosine, guanine, thymine, and uracil. The nitrogenous bases of each strand of a nucleic acid will form hydrogen bonds with certain other nitrogenous bases in a complementary strand of nucleic acid (similar to a zipper). Adenine binds with thymine and uracil; Thymine binds only with adenine; and cytosine and guanine can bind only with one another.

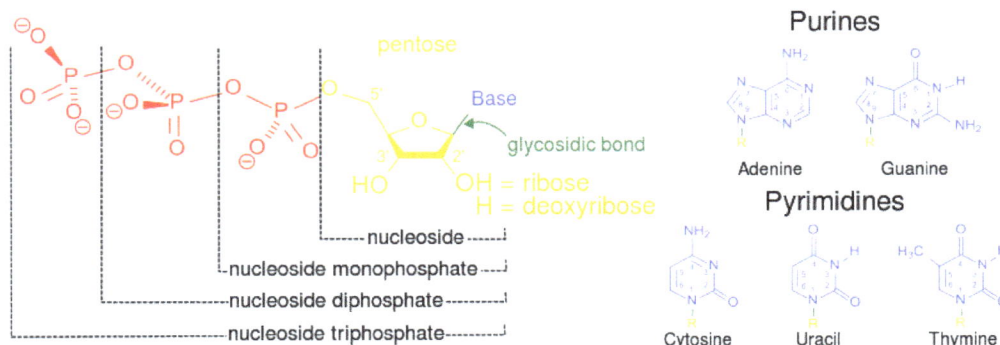

Structural elements of common nucleic acid constituents. Because they contain at least one phosphate group, the compounds marked *nucleoside monophosphate*, *nucleoside diphosphate* and *nucleoside triphosphate* are all nucleotides (not simply phosphate-lacking nucleosides).

Aside from the genetic material of the cell, nucleic acids often play a role as second messengers, as well as forming the base molecule for adenosine triphosphate (ATP), the primary energy-carrier molecule found in all living organisms. Also, the nitrogenous bases possible in the two nucleic acids are different: adenine, cytosine, and guanine occur in both RNA and DNA, while thymine occurs only in DNA and uracil occurs in RNA.

Metabolism

Carbohydrates as Energy Source

Glucose is the major energy source in most life forms. For instance, polysaccharides are broken down into their monomers (glycogen phosphorylase removes glucose residues from glycogen). Disaccharides like lactose or sucrose are cleaved into their two component monosaccharides.

Glycolysis (Anaerobic)

Glucose is mainly metabolized by a very important ten-step pathway called glycolysis, the net result of which is to break down one molecule of glucose into two molecules of pyruvate. This also produces a net two molecules of ATP, the energy currency of cells, along with two reducing equivalents of converting NAD^+ (nicotinamide adenine dinucleotide:oxidised form) to NADH (nicotinamide adenine dinucleotide:reduced form). This does not require oxygen; if no oxygen is available (or the cell cannot use oxygen), the NAD is restored by converting the pyruvate to lactate (lactic acid) (e.g., in humans) or to ethanol plus carbon dioxide (e.g., in yeast). Other monosaccharides like galactose and fructose can be converted into intermediates of the glycolytic pathway.

Aerobic

In aerobic cells with sufficient oxygen, as in most human cells, the pyruvate is further metabolized. It is irreversibly converted to acetyl-CoA, giving off one carbon atom as the waste product carbon dioxide, generating another reducing equivalent as NADH. The two molecules acetyl-CoA (from one molecule of glucose) then enter the citric acid cycle, producing two more molecules of ATP, six more NADH molecules and two reduced (ubi)quinones (via $FADH_2$ as enzyme-bound cofactor), and releasing the remaining carbon atoms as carbon dioxide. The produced NADH and quinol molecules then feed into the enzyme complexes of the respiratory chain, an electron transport system transferring the electrons ultimately to oxygen and conserving the released energy in the form of a proton gradient over a membrane (inner mitochondrial membrane in eukaryotes). Thus, oxygen is reduced to water and the original electron acceptors NAD^+ and quinone are regenerated. This is why humans breathe in oxygen and breathe out carbon dioxide. The energy released from transferring the electrons from high-energy states in NADH and quinol is conserved first as proton gradient and converted to ATP via ATP synthase. This generates an additional *28* molecules of ATP (24 from the 8 NADH + 4 from the 2 quinols), totaling to 32 molecules of ATP conserved per degraded glucose (two from glycolysis + two from the citrate cycle). It is clear that using oxygen to completely oxidize glucose provides an organism with far more energy than any oxygen-independent metabolic feature, and this is thought to be the reason why complex life appeared only after Earth's atmosphere accumulated large amounts of oxygen.

Gluconeogenesis

In vertebrates, vigorously contracting skeletal muscles (during weightlifting or sprinting, for example) do not receive enough oxygen to meet the energy demand, and so they shift to anaerobic metabolism, converting glucose to lactate. The liver regenerates the glucose, using a process called gluconeogenesis. This process is not quite the opposite of glycolysis, and actually requires three times the amount of energy gained from glycolysis (six molecules of ATP are used, compared to the two gained in glycolysis). Analogous to the above reactions, the glucose produced can then undergo glycolysis in tissues that need energy, be stored as glycogen (or starch in plants), or be converted to other monosaccharides or joined into di- or oligosaccharides. The combined pathways of glycolysis during exercise, lactate's crossing via the bloodstream to the liver, subsequent gluconeogenesis and release of glucose into the bloodstream is called the Cori cycle.

Relationship to other "Molecular-Scale" Biological Sciences

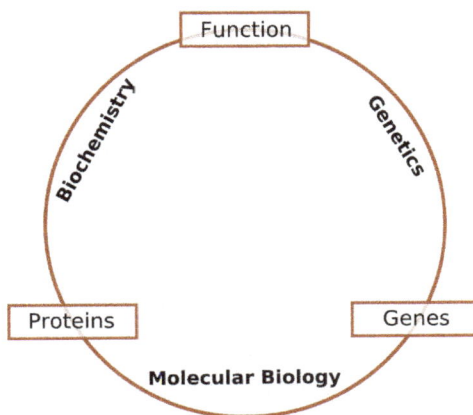

Schematic relationship between biochemistry, genetics, and molecular biology.

Researchers in biochemistry use specific techniques native to biochemistry, but increasingly combine these with techniques and ideas developed in the fields of genetics, molecular biology and biophysics. There has never been a hard-line among these disciplines in terms of content and technique. Today, the terms *molecular biology* and *biochemistry* are nearly interchangeable. The following figure is a schematic that depicts one possible view of the relationship between the fields:

- *Biochemistry* is the study of the chemical substances and vital processes occurring in living organisms. Biochemists focus heavily on the role, function, and structure of biomolecules. The study of the chemistry behind biological processes and the synthesis of biologically active molecules are examples of biochemistry.

- *Genetics* is the study of the effect of genetic differences on organisms. Often this can be inferred by the absence of a normal component (e.g., one gene). The study of "mutants" – organisms with a changed gene that leads to the organism being different with respect to the so-called "wild type" or normal phenotype. Genetic interactions (epistasis) can often confound simple interpretations of such "knock-out" or "knock-in" studies.

- *Molecular biology* is the study of molecular underpinnings of the process of replication, transcription and translation of the genetic material. The central dogma of molecular biology where genetic material is transcribed into RNA and then translated into protein, despite being an oversimplified picture of molecular biology, still provides a good starting point for understanding the field. This picture, however, is undergoing revision in light of emerging novel roles for RNA.

- *Chemical biology* seeks to develop new tools based on small molecules that allow minimal perturbation of biological systems while providing detailed information about their function. Further, chemical biology employs biological systems to create non-natural hybrids between biomolecules and synthetic devices (for example emptied viral capsids that can deliver gene therapy or drug molecules).

Nuclear Chemistry

Nuclear chemistry is the subfield of chemistry dealing with radioactivity, nuclear processes, such as nuclear transmutation, and nuclear properties.

It is the chemistry of radioactive elements such as the actinides, radium and radon together with the chemistry associated with equipment (such as nuclear reactors) which are designed to perform nuclear processes. This includes the corrosion of surfaces and the behavior under conditions of both normal and abnormal operation (such as during an accident). An important area is the behavior of objects and materials after being placed into a nuclear waste storage or disposal site.

It includes the study of the chemical effects resulting from the absorption of radiation within living animals, plants, and other materials. The radiation chemistry controls much of radiation biology as radiation has an effect on living things at the molecular scale, to explain it another way the radiation alters the biochemicals within an organism, the alteration of the biomolecules then changes the chemistry which occurs within the organism, this change in chemistry then can lead to a biological outcome. As a result, nuclear chemistry greatly assists the understanding of medical treatments (such as cancer radiotherapy) and has enabled these treatments to improve.

It includes the study of the production and use of radioactive sources for a range of processes. These include radiotherapy in medical applications; the use of radioactive tracers within industry, science and the environment; and the use of radiation to modify materials such as polymers.

It also includes the study and use of nuclear processes in *non-radioactive* areas of human activity. For instance, nuclear magnetic resonance (NMR) spectroscopy is commonly used in synthetic organic chemistry and physical chemistry and for structural analysis in macromolecular chemistry.

History

After the discovery of X-rays by Wilhelm Röntgen, many scientists began to work on ionizing radiation. One of these was Henri Becquerel, who investigated the relationship between phosphorescence and the blackening of photographic plates. When Becquerel (working in France) discovered that, with no external source of energy, the uranium generated rays which could blacken (or *fog*) the photographic plate, radioactivity was discovered. Marie Curie (working in Paris) and her husband Pierre Curie isolated two new radioactive elements from uranium ore. They used radiometric methods to identify which stream the radioactivity was in after each chemical separation; they separated the uranium ore into each of the different chemical elements that were known at the time, and measured the radioactivity of each fraction. They then attempted to separate these radioactive fractions further, to isolate a smaller fraction with a higher specific activity (radioactivity divided by mass). In this way, they isolated polonium and radium. It was noticed in about 1901 that high doses of radiation could cause an injury in humans. Henri Becquerel had carried a sample of radium in his pocket and as a result he suffered a high localised dose which resulted in a radiation burn. This injury resulted in the biological properties of radiation being investigated, which in time resulted in the development of medical treatments.

Ernest Rutherford, working in Canada and England, showed that radioactive decay can be described by a simple equation (a linear first degree derivative equation, now called first order kinetics), implying that a given radioactive substance has a characteristic "half-life" (the time taken for the amount of radioactivity present in a source to diminish by half). He also coined the terms alpha, beta and gamma rays, he converted nitrogen into oxygen, and most importantly he supervised the students who did the Geiger–Marsden experiment (gold leaf experiment) which showed that the 'plum pudding model' of the atom was wrong. In the plum pudding model, proposed by J. J. Thomson in 1904, the atom is composed of electrons surrounded by a 'cloud' of positive charge to balance the electrons' negative charge. To Rutherford, the gold foil experiment implied that the positive charge was confined to a very small nucleus leading first to the Rutherford model, and eventually to the Bohr model of the atom, where the positive nucleus is surrounded by the negative electrons.

In 1934 Marie Curie's daughter (Irène Joliot-Curie) and son-in-law (Frédéric Joliot-Curie) were the first to create artificial radioactivity: they bombarded boron with alpha particles to make the neutron-poor isotope nitrogen-13; this isotope emitted positrons. In addition, they bombarded aluminium and magnesium with neutrons to make new radioisotopes.

Main Areas

Radiochemistry is the chemistry of radioactive materials, where radioactive isotopes of elements are used to study the properties and chemical reactions of non-radioactive isotopes (often within radiochemistry the absence of radioactivity leads to a substance being described as being *inactive* as the isotopes are *stable*).

Radiation Chemistry

Radiation chemistry is the study of the chemical effects of radiation on matter; this is very different from radiochemistry as no radioactivity needs to be present in the material which is being chemically changed by the radiation. An example is the conversion of water into hydrogen gas and hydrogen peroxide.

Chemistry for Nuclear Power

Radiochemistry, radiation chemistry and nuclear chemical engineering play a very important role for uranium and thorium fuel precursors synthesis, starting from ores of these elements, fuel fabrication, coolant chemistry, fuel reprocessing, radioactive waste treatment and storage, monitoring of radioactive elements release during reactor operation and radioactive geological storage, etc.

Study of Nuclear Reactions

A combination of radiochemistry and radiation chemistry is used to study nuclear reactions such as fission and fusion. Some early evidence for nuclear fission was the formation of a short-lived radioisotope of barium which was isolated from neutron irradiated uranium (^{139}Ba, with a half-life of 83 minutes and ^{140}Ba, with a half-life of 12.8 days, are major fission products of uranium). At the time, it was thought that this was a new radium isotope, as it was then standard radiochemical practice to use a barium sulfate carrier precipitate to assist in the isolation of radium.. More recently, a combination of radiochemical methods and nuclear physics has been used to try to make new 'superheavy' elements; it is thought that islands of relative stability exist where the nuclides have half-lives of years, thus enabling weighable amounts of the new elements to be isolated. For more details of the original discovery of nuclear fission see the work of Otto Hahn.

The Nuclear Fuel Cycle

This is the chemistry associated with any part of the nuclear fuel cycle, including nuclear reprocessing. The fuel cycle includes all the operations involved in producing fuel, from mining, ore processing and enrichment to fuel production (*Front end of the cycle*). It also includes the 'in-pile' behaviour (use of the fuel in a reactor) before the *back end* of the cycle. The *back end* includes the management of the used nuclear fuel in either a spent fuel pool or dry storage, before it is disposed of into an underground waste store or reprocessed.

Normal and Abnormal Conditions

The nuclear chemistry associated with the nuclear fuel cycle can be divided into two main areas, one area is concerned with operation under the intended conditions while the other area is concerned with maloperation conditions where some alteration from the normal operating conditions has occurred or (*more rarely*) an accident is occurring.

Reprocessing

Law

In the United States it is normal to use fuel once in a power reactor before placing it in a waste store. The long term plan is currently to place the used civilian reactor fuel in a deep store. This non-reprocessing policy was started in March 1977 because of concerns about nuclear weapons proliferation. President Jimmy Carter issued a Presidential directive which indefinitely suspended the commercial reprocessing and recycling of plutonium in the United States. This directive was likely an attempt by the United States to lead other countries by example, but many other nations continue to reprocess spent nuclear fuels. The Russian government under President Vladimir Putin repealed a law which had banned the import of used nuclear fuel, which makes it possible for Russians to offer a reprocessing service for clients outside Russia (similar to that offered by BNFL).

PURex Chemistry

The current method of choice is to use the PUREX liquid-liquid extraction process which uses a tributyl phosphate/hydrocarbon mixture to extract both uranium and plutonium from nitric acid. This extraction is of the nitrate salts and is classed as being of a solvation mechanism. For example, the extraction of plutonium by an extraction agent (S) in a nitrate medium occurs by the following reaction.

$$Pu^{4+}_{aq} + 4NO_{3}^{-}{}_{aq} + 2S_{organic} \rightarrow [Pu(NO_3)_4S_2]_{organic}$$

A complex bond is formed between the metal cation, the nitrates and the tributyl phosphate, and a model compound of a dioxouranium(VI) complex with two nitrates and two triethyl phosphates has been characterised by X-ray crystallography.

When the nitric acid concentration is high the extraction into the organic phase is favoured, and when the nitric acid concentration is low the extraction is reversed (the organic phase is *stripped* of the metal). It is normal to dissolve the used fuel in nitric acid, after the removal of the insoluble matter the uranium and plutonium are extracted from the highly active liquor. It is normal to then back extract the loaded organic phase to create a *medium active* liquor which contains mostly uranium and plutonium with only small traces of fission products. This medium active aqueous mixture is then extracted again by tributyl phosphate/hydrocarbon to form a new organic phase, the metal bearing organic phase is then stripped of the metals to form an aqueous mixture of only uranium and plutonium. The two stages of extraction are used to improve the purity of the actinide product, the organic phase used for the first extraction will suffer a far greater dose of radiation. The radiation can degrade the tributyl phosphate into dibutyl hydrogen phosphate. The dibutyl hydrogen phosphate can act as an extraction agent for both the actinides and other metals such as ruthenium. The dibutyl hydrogen phosphate can make the system behave in a more complex manner as it tends to extract metals by an ion exchange mechanism (extraction favoured by low acid concentration), to reduce the effect of the dibutyl hydrogen phosphate it is common for the used organic phase to be washed with sodium carbonate solution to remove the acidic degradation products of the tributyl phosphate.

New Methods being Considered for Future Use

The PUREX process can be modified to make a UREX (*UR*anium *EX*traction) process which could be used to save space inside high level nuclear waste disposal sites, such as Yucca Mountain nuclear waste repository, by removing the uranium which makes up the vast majority of the mass and volume of used fuel and recycling it as reprocessed uranium.

The UREX process is a PUREX process which has been modified to prevent the plutonium being extracted. This can be done by adding a plutonium reductant before the first metal extraction step. In the UREX process, ~99.9% of the Uranium and >95% of Technetium are separated from each other and the other fission products and actinides. The key is the addition of acetohydroxamic acid (AHA) to the extraction and scrub sections of the process. The addition of AHA greatly diminishes the extractability of Plutonium and Neptunium, providing greater proliferation resistance than with the plutonium extraction stage of the PUREX process.

Adding a second extraction agent, octyl(phenyl)-N, N-dibutyl carbamoylmethyl phosphine oxide(CMPO) in combination with tributylphosphate, (TBP), the PUREX process can be turned into the TRUEX (*TR*ans*U*ranic *EX*traction) process this is a process which was invented in the USA by Argonne National Laboratory, and is designed to remove the transuranic metals (Am/Cm) from waste. The idea is that by lowering the alpha activity of the waste, the majority of the waste can then be disposed of with greater ease. In common with PUREX this process operates by a solvation mechanism.

As an alternative to TRUEX, an extraction process using a malondiamide has been devised. The DIAMEX (*DIAM*ide*EX*traction) process has the advantage of avoiding the formation of organic waste which contains elements other than Carbon, Hydrogen, Nitrogen, and Oxygen. Such an organic waste can be burned without the formation of acidic gases which could contribute to acid rain. The DIAMEX process is being worked on in Europe by the French CEA. The process is sufficiently mature that an industrial plant could be constructed with the existing knowledge of the process. In common with PUREX this process operates by a solvation mechanism.

Selective Actinide Extraction (SANEX). As part of the management of minor actinides it has been proposed that the lanthanides and trivalent minor actinides should be removed from the PUREX raffinate by a process such as DIAMEX or TRUEX. In order to allow the actinides such as americium to be either reused in industrial sources or used as fuel the lanthanides must be removed. The lanthanides have large neutron cross sections and hence they would poison a neutron driven nuclear reaction. To date the extraction system for the SANEX process has not been defined, but currently several different research groups are working towards a process. For instance the French CEA is working on a bis-triaiznyl pyridine (BTP) based process.

Other systems such as the dithiophosphinic acids are being worked on by some other workers.

This is the *UN*iversal *EX*traction process which was developed in Russia and the Czech Republic, it is a process designed to remove all of the most troublesome (Sr, Cs and minor actinides) radioisotopes from the raffinates left after the extraction of uranium and plutonium from used nuclear fuel. The chemistry is based upon the interaction of caesium and strontium with poly ethylene oxide (poly ethylene glycol) and a cobalt carborane anion (known as chlorinated cobalt dicarbollide) . The actinides are extracted by CMPO, and the diluent is a polar aromatic such as nitrobenzene.

Other dilents such as *meta*-nitrobenzotrifluoride and phenyl trifluoromethyl sulfone have been suggested as well.

Absorption of Fission Products on Surfaces

Another important area of nuclear chemistry is the study of how fission products interact with surfaces; this is thought to control the rate of release and migration of fission products both from waste containers under normal conditions and from power reactors under accident conditions. It is interesting to note that, like chromate and molybdate, the $^{99}TcO_4$ anion can react with steel surfaces to form a corrosion resistant layer. In this way, these metaloxo anions act as anodic corrosion inhibitors. The formation of $^{99}TcO_2$ on steel surfaces is one effect which will retard the release of ^{99}Tc from nuclear waste drums and nuclear equipment which has been lost before decontamination (e.g. submarine reactors lost at sea). This $^{99}TcO_2$ layer renders the steel surface passive, inhibiting the anodic corrosion reaction. The radioactive nature of technetium makes this corrosion protection impractical in almost all situations. It has also been shown that $^{99}TcO_4$ anions react to form a layer on the surface of activated carbon (charcoal) or aluminium.. A short review of the biochemical properties of a series of key long lived radioisotopes can be read on line.

^{99}Tc in nuclear waste may exist in chemical forms other than the $^{99}TcO_4$ anion, these other forms have different chemical properties.

Similarly, the release of iodine-131 in a serious power reactor accident could be retarded by absorption on metal surfaces within the nuclear plant.

Education

Despite the growing use of nuclear medicine, the potential expansion of nuclear power plants, and worries about protection against nuclear threats and the management of the nuclear waste generated in past decades, the number of students opting to specialize in nuclear and radiochemistry has decreased significantly over the past few decades. Now, with many experts in these fields approaching retirement age, action is needed to avoid a workforce gap in these critical fields, for example by building student interest in these careers, expanding the educational capacity of universities and colleges, and providing more specific on-the-job training.

Nuclear and Radiochemistry (NRC) is mostly being taught at university level, usually first at the Master- and PhD-degree level. In Europe, as substantial effort is being done to harmonize and prepare the NRC education for the industry's and society's future needs. This effort is being coordinated in a projects funded by the Coordinated Action supported by the European Atomic Energy Community's 7th Framework Program: The CINCH-II project - Cooperation in education and training In Nuclear Chemistry. This project has set up a wiki dedicated to NRC teaching: NucWik. Although NucWik is primarily aimed at teachers, anyone interested in Nuclear and Radiochemistry is welcome and can find a lot of information and material explaining topics related to NRC.

Spinout Areas

Some methods first developed within nuclear chemistry and physics have become so widely used within chemistry and other physical sciences that they may be best thought of as separate from *normal* nuclear chemistry. For example, the isotope effect is used so extensively to investigate chemical mechanisms and the use of cosmogenic isotopes and long-lived unstable isotopes in geology that it is best to consider much of isotopic chemistry as separate from nuclear chemistry.

Kinetics (Use within Mechanistic Chemistry)

The mechanisms of chemical reactions can be investigated by observing how the kinetics of a reaction is changed by making an isotopic modification of a substrate, known as the kinetic isotope effect. This is now a standard method in organic chemistry. Briefly, replacing normal hydrogen (protons) by deuterium within a molecule causes the molecular vibrational frequency of X-H (for example C-H, N-H and O-H) bonds to decrease, which leads to a decrease in vibrational zero-point energy. This can lead to a decrease in the reaction rate if the rate-determining step involves breaking a bond between hydrogen and another atom. Thus, if the reaction changes in rate when protons are replaced by deuteriums, it is reasonable to assume that the breaking of the bond to hydrogen is part of the step which determines the rate.

Uses within Geology, Biology and Forensic Science

Cosmogenic isotopes are formed by the interaction of cosmic rays with the nucleus of an atom. These can be used for dating purposes and for use as natural tracers. In addition, by careful measurement of some ratios of stable isotopes it is possible to obtain new insights into the origin of bullets, ages of ice samples, ages of rocks, and the diet of a person can be identified from a hair or other tissue sample.

Biology

Within living things, isotopic labels (both radioactive and nonradioactive) can be used to probe how the complex web of reactions which makes up the metabolism of an organism converts one substance to another. For instance a green plant uses light energy to convert water and carbon dioxide into glucose by photosynthesis. If the oxygen in the water is labeled, then the label appears in the oxygen gas formed by the plant and not in the glucose formed in the chloroplasts within the plant cells.

For biochemical and physiological experiments and medical methods, a number of specific isotopes have important applications.

- Stable isotopes have the advantage of not delivering a radiation dose to the system being studied; however, a significant excess of them in the organ or organism might still interfere with its functionality, and the availability of sufficient amounts for whole-animal studies is limited for many isotopes. Measurement is also difficult, and usually requires mass spectrometry to determine how much of the isotope is present in particular compounds, and there is no means of localizing measurements within the cell.

- H-2 (deuterium), the stable isotope of hydrogen, is a stable tracer, the concentration of which can be measured by mass spectrometry or NMR. It is incorporated into all cellular structures. Specific deuterated compounds can also be produced.

- N-15, the stable isotope of nitrogen, has also been used. It is incorporated mainly into proteins.

- Radioactive isotopes have the advantages of being detectable in very low quantities, in being easily measured by scintillation counting or other radiochemical methods, and in being localizable to particular regions of a cell, and quantifiable by autoradiography. Many compounds with the radioactive atoms in specific positions can be prepared, and are widely available commercially. In high quantities they require precautions to guard the workers from the effects of radiation—and they can easily contaminate laboratory glassware and other equipment. For some isotopes the half-life is so short that preparation and measurement is difficult.

By organic synthesis it is possible to create a complex molecule with a radioactive label that can be confined to a small area of the molecule. For short-lived isotopes such as ^{11}C, very rapid synthetic methods have been developed to permit the rapid addition of the radioactive isotope to the molecule. For instance a palladium catalysed carbonylation reaction in a microfluidic device has been used to rapidly form amides and it might be possible to use this method to form radioactive imaging agents for PET imaging.

- ^3H, Tritium, the radioisotope of hydrogen, it available at very high specific activities, and compounds with this isotope in particular positions are easily prepared by standard chemical reactions such as hydrogenation of unsaturated precursors. The isotope emits very soft beta radiation, and can be detected by scintillation counting.

- ^{11}C, Carbon-11 is usually produced by cyclotron bombardment of ^{14}N with protons. The resulting nuclear reaction is $^{14}N(p,\alpha)^{11}C$. Additionally, Carbon-11 can also be made using a cyclotron; boron in the form of boric oxide is reacted with protons in a (p,n) reaction. Another alternative route is to react ^{10}B with deuterons. By rapid organic synthesis, the ^{11}C compound formed in the cyclotron is converted into the imaging agent which is then used for PET.

- ^{14}C, Carbon-14 can be made (as above), and it is possible to convert the target material into simple inorganic and organic compounds. In most organic synthesis work it is normal to try to create a product out of two approximately equal sized fragments and to use a convergent route, but when a radioactive label is added, it is normal to try to add the label late in the synthesis in the form of a very small fragment to the molecule to enable the radioactivity to be localised in a single group. Late addition of the label also reduces the number of synthetic stages where radioactive material is used.

- ^{18}F, fluorine-18 can be made by the reaction of neon with deuterons, ^{20}Ne reacts in a (d,^4He) reaction. It is normal to use neon gas with a trace of stable fluorine ($^{19}F_2$). The $^{19}F_2$ acts as a carrier which increases the yield of radioactivity from the cyclotron target by reducing the amount of radioactivity lost by absorption on surfaces. However, this reduction in loss is at the cost of the specific activity of the final product.

Nuclear Magnetic Resonance (NMR)

NMR spectroscopy uses the net spin of nuclei in a substance upon energy absorption to identify molecules. This has now become a standard spectroscopic tool within synthetic chemistry. One major use of NMR is to determine the bond connectivity within an organic molecule.

NMR imaging also uses the net spin of nuclei (commonly protons) for imaging. This is widely used for diagnostic purposes in medicine, and can provide detailed images of the inside of a person without inflicting any radiation upon them. In a medical setting, NMR is often known simply as "magnetic resonance" imaging, as the word 'nuclear' has negative connotations for many people.

Physical Chemistry

Physical chemistry is the study of macroscopic, atomic, subatomic, and particulate phenomena in chemical systems in terms of laws and concepts of physics. It applies the principles, practices and concepts of physics such as motion, energy, force, time, thermodynamics, quantum chemistry, statistical mechanics, dynamics and equilibrium.

Physical chemistry, in contrast to chemical physics, is predominantly (but not always) a macroscopic or supra-molecular science, as the majority of the principles on which physical chemistry was founded are concepts related to the bulk rather than on molecular/atomic structure alone (for example, chemical equilibrium and colloids).

Some of the relationships that physical chemistry strives to resolve include the effects of:

1. Intermolecular forces that act upon the physical properties of materials (plasticity, tensile strength, surface tension in liquids).

2. Reaction kinetics on the rate of a reaction.

3. The identity of ions and the electrical conductivity of materials.

4. Surface chemistry and electrochemistry of membranes.

5. Interaction of one body with another in terms of quantities of heat and work called thermodynamics.

6. Transfer of heat between a chemical system and its surroundings during change of phase or chemical reaction taking place called thermochemistry

7. Study of colligative properties of number of species present in solution.

8. Number of phases, number of components and degree of freedom (or variance) can be correlated with one another with help of phase rule.

9. Reactions of electrochemical cells.

Key Concepts

The key concepts of physical chemistry are the ways in which pure physics is applied to chemical problems.

One of the key concepts in classical chemistry is that all chemical compounds can be described as groups of atoms bonded together and chemical reactions can be described as the making and breaking of those bonds. Predicting the properties of chemical compounds from a description of atoms and how they bond is one of the major goals of physical chemistry. To describe the atoms and bonds precisely, it is necessary to know both where the nuclei of the atoms are, and how electrons are distributed around them. Quantum chemistry, a subfield of physical chemistry especially concerned with the application of quantum mechanics to chemical problems, provides tools to determine how strong and what shape bonds are, how nuclei move, and how light can be absorbed or emitted by a chemical compound. Spectroscopy is the related sub-discipline of physical chemistry which is specifically concerned with the interaction of electromagnetic radiation with matter.

Another set of important questions in chemistry concerns what kind of reactions can happen spontaneously and which properties are possible for a given chemical mixture. This is studied in chemical thermodynamics, which sets limits on quantities like how far a reaction can proceed, or how much energy can be converted into work in an internal combustion engine, and which provides links between properties like the thermal expansion coefficient and rate of change of entropy with pressure for a gas or a liquid. It can frequently be used to assess whether a reactor or engine design is feasible, or to check the validity of experimental data. To a limited extent, quasi-equilibrium and non-equilibrium thermodynamics can describe irreversible changes. However, classical thermodynamics is mostly concerned with systems in equilibrium and reversible changes and not what actually does happen, or how fast, away from equilibrium.

Which reactions do occur and how fast is the subject of chemical kinetics, another branch of physical chemistry. A key idea in chemical kinetics is that for reactants to react and form products, most chemical species must go through transition states which are higher in energy than either the reactants or the products and serve as a barrier to reaction. In general, the higher the barrier, the slower the reaction. A second is that most chemical reactions occur as a sequence of elementary reactions, each with its own transition state. Key questions in kinetics include how the rate of reaction depends on temperature and on the concentrations of reactants and catalysts in the reaction mixture, as well as how catalysts and reaction conditions can be engineered to optimize the reaction rate.

The fact that how fast reactions occur can often be specified with just a few concentrations and a temperature, instead of needing to know all the positions and speeds of every molecule in a mixture, is a special case of another key concept in physical chemistry, which is that to the extent an engineer needs to know, everything going on in a mixture of very large numbers (perhaps of the order of the Avogadro constant, 6×10^{23}) of particles can often be described by just a few variables like pressure, temperature, and concentration. The precise reasons for this are described in statistical mechanics, a specialty within physical chemistry which is also shared with physics. Statistical mechanics also provides ways to predict the properties we see in everyday life from molecular properties without relying on empirical correlations based on chemical similarities.

History

The term "physical chemistry" was coined by Mikhail Lomonosov in 1752, when he presented a lecture course entitled "A Course in True Physical Chemistry" (Russian: «Курс истинной физической химии») before the students of Petersburg University. In the preamble to these lectures he gives definition: "Physical chemistry is the science that must explain under provisions of physical experiments the reason for what is happening in complex bodies through chemical operations".

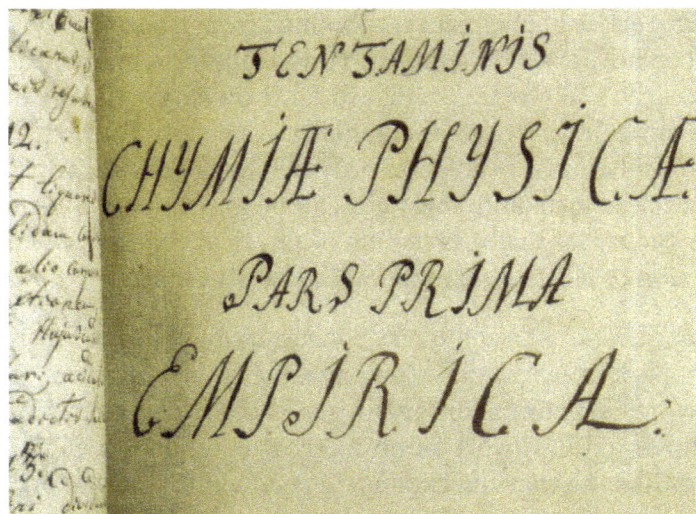

Fragment of M. Lomonosov's manuscript 'Physical Chemistry' (1752)

Modern physical chemistry originated in the 1860s to 1880s with work on chemical thermodynamics, electrolytes in solutions, chemical kinetics and other subjects. One milestone was the publication in 1876 by Josiah Willard Gibbs of his paper, *On the Equilibrium of Heterogeneous Substances*. This paper introduced several of the cornerstones of physical chemistry, such as Gibbs energy, chemical potentials, and Gibbs' phase rule. Other milestones include the subsequent naming and accreditation of enthalpy to Heike Kamerlingh Onnes and to macromolecular processes.

The first scientific journal specifically in the field of physical chemistry was the German journal, *Zeitschrift für Physikalische Chemie*, founded in 1887 by Wilhelm Ostwald and Jacobus Henricus van 't Hoff. Together with Svante August Arrhenius, these were the leading figures in physical chemistry in the late 19th century and early 20th century. All three were awarded with the Nobel Prize in Chemistry between 1901-1909.

Developments in the following decades include the application of statistical mechanics to chemical systems and work on colloids and surface chemistry, where Irving Langmuir made many contributions. Another important step was the development of quantum mechanics into quantum chemistry from the 1930s, where Linus Pauling was one of the leading names. Theoretical developments have gone hand in hand with developments in experimental methods, where the use of different forms of spectroscopy, such as infrared spectroscopy, microwave spectroscopy, EPR spectroscopy and NMR spectroscopy, is probably the most important 20th century development.

Further development in physical chemistry may be attributed to discoveries in nuclear chemistry, especially in isotope separation (before and during World War II), more recent discoveries in astrochemistry, as well as the development of calculation algorithms in the field of "additive physicochemical properties" (practically all physicochemical properties, such as boiling point, critical point, surface tension, vapor pressure, etc. - more than 20 in all - can be precisely calculated from chemical structure alone, even if the chemical molecule remains unsynthesized), and in this area is concentrated practical importance of contemporary physical chemistry.

See Group contribution method, Lydersen method, Joback method, Benson group increment theory, QSPR, QSAR

Astrochemistry

Astrochemistry is the study of the abundance and reactions of chemical elements and molecules in the universe, and their interaction with radiation. The discipline is an overlap of astronomy and chemistry. The word "astrochemistry" may be applied to both the Solar System and the interstellar medium. The study of the abundance of elements and isotope ratios in Solar System objects, such as meteorites, is also called cosmochemistry, while the study of interstellar atoms and molecules and their interaction with radiation is sometimes called molecular astrophysics. The formation, atomic and chemical composition, evolution and fate of molecular gas clouds is of special interest, because it is from these clouds that solar systems form.

Spectroscopy

One particularly important experimental tool in astrochemistry is spectroscopy, the use of telescopes to measure the absorption and emission of light from molecules and atoms in various environments. By comparing astronomical observations with laboratory measurements, astrochemists can infer the elemental abundances, chemical composition, and temperatures of stars and interstellar clouds. This is possible because ions, atoms, and molecules have characteristic spectra: that is, the absorption and emission of certain wavelengths (colors) of light, often not visible to the human eye. However, these measurements have limitations, with various types of radiation (radio, infrared, visible, ultraviolet etc.) able to detect only certain types of species, depending on the chemical properties of the molecules. Interstellar formaldehyde was the first organic molecule detected in the interstellar medium.

Perhaps the most powerful technique for detection of individual chemical species is radio astronomy, which has resulted in the detection of over a hundred interstellar species, including radicals and ions, and organic (i.e. carbon-based) compounds, such as alcohols, acids, aldehydes, and ketones. One of the most abundant interstellar molecules, and among the easiest to detect with radio waves (due to its strong electric dipole moment), is CO (carbon monoxide). In fact, CO is such a common interstellar molecule that it is used to map out molecular regions. The radio observation of perhaps greatest human interest is the claim of interstellar glycine, the simplest amino acid, but with considerable accompanying controversy. One of the reasons why this detection was controversial is that although radio (and some other methods like rotational spectroscopy) are good for the identification of simple species with large dipole moments, they are less sensitive to more complex molecules, even something relatively small like amino acids.

Moreover, such methods are completely blind to molecules that have no dipole. For example, by far the most common molecule in the universe is H_2 (hydrogen gas), but it does not have a dipole moment, so it is invisible to radio telescopes. Moreover, such methods cannot detect species that are not in the gas-phase. Since dense molecular clouds are very cold (10 to 50 K [−263.1 to −223.2 °C; −441.7 to −369.7 °F]), most molecules in them (other than hydrogen) are frozen, i.e. solid. Instead, hydrogen and these other molecules are detected using other wavelengths of light. Hydrogen is easily detected in the ultraviolet (UV) and visible ranges from its absorption and emission of light (the hydrogen line). Moreover, most organic compounds absorb and emit light in the infrared (IR) so, for example, the detection of methane in the atmosphere of Mars was achieved using an IR ground-based telescope, NASA's 3-meter Infrared Telescope Facility atop Mauna Kea, Hawaii. NASA also has an airborne IR telescope called SOFIA and an IR space telescope called Spitzer. Somewhat related to the recent detection of methane in the atmosphere of Mars, scientists reported, in June 2012, that measuring the ratio of hydrogen and methane levels on Mars may help determine the likelihood of life on Mars. According to the scientists, "...low H_2/CH_4 ratios (less than approximately 40) indicate that life is likely present and active." Other scientists have recently reported methods of detecting hydrogen and methane in extraterrestrial atmospheres.

Infrared astronomy has also revealed that the interstellar medium contains a suite of complex gas-phase carbon compounds called polyaromatic hydrocarbons, often abbreviated PAHs or PACs. These molecules, composed primarily of fused rings of carbon (either neutral or in an ionized state), are said to be the most common class of carbon compound in the galaxy. They are also the most common class of carbon molecule in meteorites and in cometary and asteroidal dust (cosmic dust). These compounds, as well as the amino acids, nucleobases, and many other compounds in meteorites, carry deuterium and isotopes of carbon, nitrogen, and oxygen that are very rare on earth, attesting to their extraterrestrial origin. The PAHs are thought to form in hot circumstellar environments (around dying, carbon-rich red giant stars).

Infrared astronomy has also been used to assess the composition of solid materials in the interstellar medium, including silicates, kerogen-like carbon-rich solids, and ices. This is because unlike visible light, which is scattered or absorbed by solid particles, the IR radiation can pass through the microscopic interstellar particles, but in the process there are absorptions at certain wavelengths that are characteristic of the composition of the grains. As above with radio astronomy, there are certain limitations, e.g. N_2 is difficult to detect by either IR or radio astronomy.

Such IR observations have determined that in dense clouds (where there are enough particles to attenuate the destructive UV radiation) thin ice layers coat the microscopic particles, permitting some low-temperature chemistry to occur. Since hydrogen is by far the most abundant molecule in the universe, the initial chemistry of these ices is determined by the chemistry of the hydrogen. If the hydrogen is atomic, then the H atoms react with available O, C and N atoms, producing "reduced" species like H_2O, CH_4, and NH_3. However, if the hydrogen is molecular and thus not reactive, this permits the heavier atoms to react or remain bonded together, producing CO, CO_2, CN, etc. These mixed-molecular ices are exposed to ultraviolet radiation and cosmic rays, which results in complex radiation-driven chemistry. Lab experiments on the photochemistry of simple interstellar ices have produced amino acids. The similarity between interstellar and cometary ices (as well as comparisons of gas phase compounds) have been invoked as indicators of a connection between interstellar and cometary chemistry. This is somewhat supported by the results of the analysis of the organics from the comet samples returned by the Stardust mission but the minerals also indicated a surprising contribution from high-temperature chemistry in the solar nebula.

Research

Research is progressing on the way in which interstellar and circumstellar molecules form and interact, and this research could have a profound impact on our understanding of the suite of molecules that were present in the molecular cloud when our solar system formed, which contributed to the rich carbon chemistry of comets and asteroids and hence the meteorites and interstellar dust particles which fall to the Earth by the ton every day.

Transition from atomic to molecular gas at the border of the Orion molecular cloud.

The sparseness of interstellar and interplanetary space results in some unusual chemistry, since symmetry-forbidden reactions cannot occur except on the longest of timescales. For this reason, molecules and molecular ions which are unstable on Earth can be highly abundant in space, for example the H_3^+ ion. Astrochemistry overlaps with astrophysics and nuclear physics in characterizing the nuclear reactions which occur in stars, the consequences for stellar evolution, as well as stellar 'generations'. Indeed, the nuclear reactions in stars produce every naturally occurring chemical element. As the stellar 'generations' advance, the mass of the newly formed elements increases. A first-generation star uses elemental hydrogen (H) as a fuel source and produces helium (He). Hydrogen is the most abundant element, and it is the basic building block for all other elements as its nucleus has only one proton. Gravitational pull toward the center of a star creates massive amounts of heat and pressure, which cause nuclear fusion. Through this process of merging nuclear mass, heavier elements are formed. Carbon, oxygen and silicon are examples of elements that form in stellar fusion. After many stellar generations, very heavy elements are formed (e.g. iron and lead).

In October 2011, scientists reported that cosmic dust contains organic matter ("amorphous organic solids with a mixed aromatic-aliphatic structure") that could be created naturally, and rapidly, by stars.

On August 29, 2012, and in a world first, astronomers at Copenhagen University reported the detection of a specific sugar molecule, glycolaldehyde, in a distant star system. The molecule was found around the protostellar binary *IRAS 16293-2422*, which is located 400 light years from Earth. Glycolaldehyde is needed to form ribonucleic acid, or RNA, which is similar in function to DNA. This finding suggests that complex organic molecules may form in stellar systems prior to the formation of planets, eventually arriving on young planets early in their formation.

In September, 2012, NASA scientists reported that polycyclic aromatic hydrocarbons (PAHs), subjected to interstellar medium (ISM) conditions, are transformed, through hydrogenation, oxygenation and hydroxylation, to more complex organics - "a step along the path toward amino acids and nucleotides, the raw materials of proteins and DNA, respectively". Further, as a result of these transformations, the PAHs lose their spectroscopic signature which could be one of the reasons "for the lack of PAH detection in interstellar ice grains, particularly the outer regions of cold, dense clouds or the upper molecular layers of protoplanetary disks."

In February 2014, NASA announced the creation of an improved spectral database for tracking polycyclic aromatic hydrocarbons (PAHs) in the universe. According to scientists, more than 20% of the carbon in the universe may be associated with PAHs, possible starting materials for the formation of life. PAHs seem to have been formed shortly after the Big Bang, are widespread throughout the universe, and are associated with new stars and exoplanets.

On August 11, 2014, astronomers released studies, using the Atacama Large Millimeter/Submillimeter Array (ALMA) for the first time, that detailed the distribution of HCN, HNC, H_2CO, and dust inside the comae of comets C/2012 F6 (Lemmon) and C/2012 S1 (ISON).

For the study of the recourses of chemical elements and molecules in the universe is developed the mathematical model of the molecules composition distribution in the interstellar environment on thermodynamic potentials by professor M.Yu. Dolomatov using methods of the probability theory, the mathematical and physical statistics and the equilibrium thermodynamics. Based on this model are estimated the resources of life-related molecules, amino acids and the nitrogenous bases in the interstellar medium. The possibility of the oil hydrocarbons molecules formation is shown. The given calculations confirm Sokolov's and Hoyl's hypotheses about the possibility of the oil hydrocarbons formation in Space. Results are confirmed by data of astrophysical supervision and space researches.

In July 2015, scientists reported that upon the first touchdown of the *Philae* lander on comet 67/P's surface, measurements by the COSAC and Ptolemy instruments revealed sixteen organic compounds, four of which were seen for the first time on a comet, including acetamide, acetone, methyl isocyanate and propionaldehyde.

Organic Chemistry

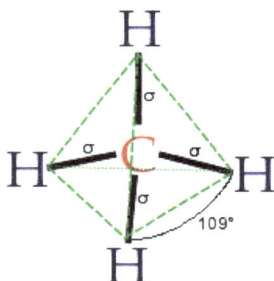

Methane, CH_4, in line-angle representation showing four carbon-hydrogen single (σ) bonds in black, and the 3D shape of such tetrahedral molecules, with ~109° interior bond angles, in green. Methane is the simplest organic chemical and simplest hydrocarbon, and molecules can be built up conceptually from it by exchanging up to all 4 hydrogens with carbon or other atoms.

Organic chemistry is a chemistry subdiscipline involving the scientific study of the structure, properties, and reactions of organic compounds and organic materials, i.e., matter in its various forms that contain carbon atoms. Study of structure includes many physical and chemical methods to determine the chemical composition and the chemical constitution of organic compounds and materials. Study of properties includes both physical properties and chemical properties, and uses similar methods as well as methods to evaluate chemical reactivity, with the aim to understand the behavior of the organic matter in its pure form (when possible), but also in solutions, mixtures, and fabricated forms. The study of organic reactions includes probing their scope through use in preparation of target compounds (e.g., natural products, drugs, polymers, etc.) by chemical synthesis, as well as the focused study of the reactivities of individual organic molecules, both in the laboratory and via theoretical (in silico) study.

The range of chemicals studied in organic chemistry include hydrocarbons (compounds containing only carbon and hydrogen), as well as myriad compositions based always on carbon, but also containing other elements, especially oxygen, nitrogen, sulfur, phosphorus (these, included in many organic chemicals in biology) and the radiostable elements of the halogens.

In the modern era, the range extends further into the periodic table, with main group elements, including:

- Group 1 and 2 organometallic compounds, i.e., involving alkali (e.g., lithium, sodium, and potassium) or alkaline earth metals (e.g., magnesium)

- Metalloids (e.g., boron and silicon) or other metals (e.g., aluminium and tin)

In addition, much modern research focuses on organic chemistry involving further organometallics, including the lanthanides, but especially the transition metals; (e.g., zinc, copper, palladium, nickel, cobalt, titanium and chromium)

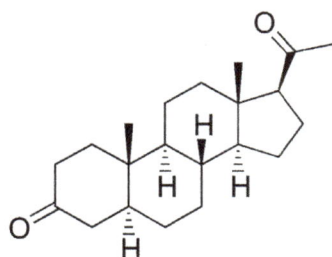

Line-angle representation   Ball-and-stick representation   Space-filling representation

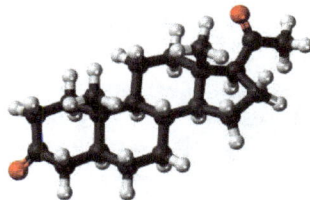

Three representations of an organic compound, 5α-Dihydroprogesterone (5α-DHP), a steroid hormone. For molecules showing color, the carbon atoms are in black, hydrogens in gray, and oxygens in red. In the line angle representation, carbon atoms are implied at every terminus of a line and vertex of multiple lines, and hydrogen atoms are implied to fill the remaining needed valences (up to 4).

Finally, organic compounds form the basis of all earthly life and constitute a significant part of human endeavors in chemistry. The bonding patterns open to carbon, with its valence of four—formal single, double, and triple bonds, as well as various structures with delocalized electrons—make the array of organic compounds structurally diverse, and their range of applications enormous. They either form the basis of, or are important constituents of, many commercial products including pharmaceuticals; petrochemicals and products made from them (including lubricants, solvents, etc.); plastics; fuels and explosives; etc. As indicated, the study of organic chemistry overlaps with organometallic chemistry and biochemistry, but also with medicinal chemistry, polymer chemistry, as well as many aspects of materials science.

Periodic table of elements of interest in organic chemistry. The table illustrates all elements of current interest in modern organic and organometallic chemistry, indicating main group elements in orange, and transition metals and lanthanides (Lan) in grey.

History

Friedrich Wöhler

Before the nineteenth century, chemists generally believed that compounds obtained from living organisms were endowed with a vital force that distinguished them from inorganic compounds. According to the concept of vitalism (vital force theory), organic matter was endowed with a "vital force". During the first half of the nineteenth century, some of the first systematic studies of organic compounds were reported. Around 1816 Michel Chevreul started a study of soaps made from various fats and alkalis. He separated the different acids that, in combination with the alkali, produced the soap. Since these were all individual compounds, he demonstrated that it was possible to make a chemical change in various fats (which traditionally come from organic sources), producing new compounds, without "vital force". In 1828 Friedrich Wöhler produced the organic chemical urea (carbamide), a constituent of urine, from the inorganic ammonium cyanate NH_4CNO, in what is now called the Wöhler synthesis. Although Wöhler was always cautious about claiming that he had disproved the theory of vital force, this event has often been thought of as a turning point.

In 1856 William Henry Perkin, while trying to manufacture quinine, accidentally manufactured the organic dye now known as Perkin's mauve. Through its great financial success, this discovery greatly increased interest in organic chemistry.

The crucial breakthrough for organic chemistry was the concept of chemical structure, developed independently and simultaneously by Friedrich August Kekulé and Archibald Scott Couper in 1858. Both men suggested that tetravalent carbon atoms could link to each other to form a carbon lattice, and that the detailed patterns of atomic bonding could be discerned by skillful interpretations of appropriate chemical reactions.

The pharmaceutical industry began in the last decade of the 19th century when the manufacturing of acetylsalicylic acid—more commonly referred to as aspirin—in Germany was started by Bayer. The first time a drug was systematically improved was with arsphenamine (Salvarsan). Though numerous derivatives of the dangerous toxic atoxyl were examined by Paul Ehrlich and his group, the compound with best effectiveness and toxicity characteristics was selected for production.

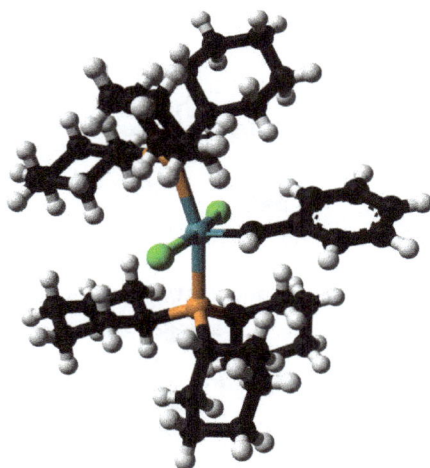

An example of an organometallic molecule, a catalyst called Grubbs' catalyst, as a ball-and-stick model based on an X-ray crystal structure. The formula of the catalyst is often given as $RuCl_2(PCy_3)_2(=CHPh)$, where the ruthenium metal atom, Ru, is at very center in turquoise, carbons are in black, hydrogens in gray-white, chlorine in green, and phosphorus in orange. The metal ligand at the bottom is a tricyclohexyl phosphine, abbreviated PCy, and another of these appears at the top of the image (where its rings are obscuring one another). The group projecting out to the right has a metal-carbon double bond, as is known as an alkylidene. Robert Grubbs shared the 2005 Nobel prize in chemistry with Richard R. Schrock and Yves Chauvin for their work on the reactions such catalysts mediate, called olefin metathesis.

Early examples of organic reactions and applications were often found because of a combination of luck and preparation for unexpected observations. The latter half of the 19th century however witnessed systematic studies of organic compounds. The development of synthetic indigo is illustrative. The production of indigo from plant sources dropped from 19,000 tons in 1897 to 1,000 tons by 1914 thanks to the synthetic methods developed by Adolf von Baeyer. In 2002, 17,000 tons of synthetic indigo were produced from petrochemicals.

In the early part of the 20th Century, polymers and enzymes were shown to be large organic molecules, and petroleum was shown to be of biological origin.

The multiple-step synthesis of complex organic compounds is called total synthesis. Total synthesis of complex natural compounds increased in complexity to glucose and terpineol. For example, cholesterol-related compounds have opened ways to synthesize complex human hormones and their modified derivatives. Since the start of the 20th century, complexity of total syntheses has been increased to include molecules of high complexity such as lysergic acid and vitamin B12.

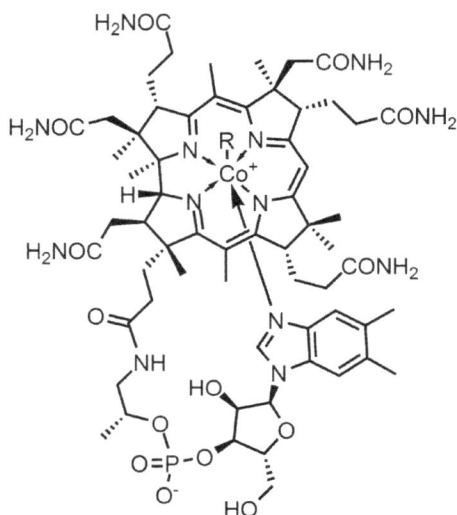

R = 5'-deoxyadenosyl, Me, OH, CN

The total synthesis of vitamin B12 marked a major achievement in organic chemistry.

The development of organic chemistry benefited from the discovery of petroleum and the development of the petrochemical industry. The conversion of individual compounds obtained from petroleum into different compound types by various chemical processes led to the birth of the petrochemical industry, which successfully manufactured artificial rubbers, various organic adhesives, property-modifying petroleum additives, and plastics.

The majority of chemical compounds occurring in biological organisms are in fact carbon compounds, so the association between organic chemistry and biochemistry is so close that biochemistry might be regarded as in essence a branch of organic chemistry. Although the history of biochemistry might be taken to span some four centuries, fundamental understanding of the field only began to develop in the late 19th century and the actual term *biochemistry* was coined around the start of 20th century. Research in the field increased throughout the twentieth century, without any indication of slackening in the rate of increase, as may be verified by inspection of abstraction and indexing services such as BIOSIS Previews and Biological Abstracts, which began in the 1920s as a single annual volume, but has grown so drastically that by the end of the 20th century it was only available to the everyday user as an online electronic database.

Characterization

Since organic compounds often exist as mixtures, a variety of techniques have also been developed to assess purity, especially important being chromatography techniques such as HPLC and gas chromatography. Traditional methods of separation include distillation, crystallization, and solvent extraction.

Organic compounds were traditionally characterized by a variety of chemical tests, called "wet methods", but such tests have been largely displaced by spectroscopic or other computer-intensive methods of analysis. Listed in approximate order of utility, the chief analytical methods are:

- Nuclear magnetic resonance (NMR) spectroscopy is the most commonly used technique, often permitting complete assignment of atom connectivity and even stereochemistry using correlation spectroscopy. The principal constituent atoms of organic chemistry – hydrogen and carbon – exist naturally with NMR-responsive isotopes, respectively ^{1}H and ^{13}C.

- Elemental analysis: A destructive method used to determine the elemental composition of a molecule.

- Mass spectrometry indicates the molecular weight of a compound and, from the fragmentation patterns, its structure. High resolution mass spectrometry can usually identify the exact formula of a compound and is used in lieu of elemental analysis. In former times, mass spectrometry was restricted to neutral molecules exhibiting some volatility, but advanced ionization techniques allow one to obtain the "mass spec" of virtually any organic compound.

- Crystallography is an unambiguous method for determining molecular geometry, the proviso being that single crystals of the material must be available and the crystal must be representative of the sample. Highly automated software allows a structure to be determined within hours of obtaining a suitable crystal.

Traditional spectroscopic methods such as infrared spectroscopy, optical rotation, UV/VIS spectroscopy provide relatively nonspecific structural information but remain in use for specific classes of compounds. Traditionally refractive index and density were also important for substance identification.

Properties

Physical properties of organic compounds typically of interest include both quantitative and qualitative features. Quantitative information includes melting point, boiling point, and index of refraction. Qualitative properties include odor, consistency, solubility, and color.

Melting and Boiling Properties

Organic compounds typically melt and many boil. In contrast, while inorganic materials generally can be melted, many do not boil, tending instead to degrade. In earlier times, the melting point (m.p.) and boiling point (b.p.) provided crucial information on the purity and identity of organic compounds. The melting and boiling points correlate with the polarity of the molecules and their molecular weight. Some organic compounds, especially symmetrical ones, sublime, that is they evaporate without melting. A well-known example of a sublimable organic compound is para-di-chlorobenzene, the odiferous constituent of modern mothballs. Organic compounds are usually not very stable at temperatures above 300 °C, although some exceptions exist.

Solubility

Neutral organic compounds tend to be hydrophobic; that is, they are less soluble in water than in organic solvents. Exceptions include organic compounds that contain ionizable (which can be converted in ions) groups as well as low molecular weight alcohols, amines, and carboxylic acids where hydrogen bonding occurs. Organic compounds tend to dissolve in organic solvents. Solvents can be either pure substances like ether or ethyl alcohol, or mixtures, such as the paraffinic solvents such as the various petroleum ethers and white spirits, or the range of pure or mixed aromatic solvents obtained from petroleum or tar fractions by physical separation or by chemical conversion. Solubility in the different solvents depends upon the solvent type and on the functional groups if present.

Solid State Properties

Various specialized properties of molecular crystals and organic polymers with conjugated systems are of interest depending on applications, e.g. thermo-mechanical and electro-mechanical such as piezoelectricity, electrical conductivity and electro-optical (e.g. non-linear optics) properties. For historical reasons, such properties are mainly the subjects of the areas of polymer science and materials science.

Nomenclature

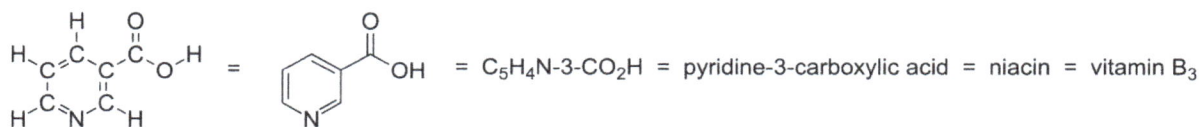

Various names and depictions for one organic compound.

The names of organic compounds are either systematic, following logically from a set of rules, or nonsystematic, following various traditions. Systematic nomenclature is stipulated by specifications from IUPAC. Systematic nomenclature starts with the name for a parent structure within the molecule of interest. This parent name is then modified by prefixes, suffixes, and numbers to unambiguously convey the structure. Given that millions of organic compounds are known, rigorous use of systematic names can be cumbersome. Thus, IUPAC recommendations are more closely followed for simple compounds, but not complex molecules. To use the systematic naming, one must know the structures and names of the parent structures. Parent structures include unsubstituted hydrocarbons, heterocycles, and monofunctionalized derivatives thereof.

Nonsystematic nomenclature is simpler and unambiguous, at least to organic chemists. Nonsystematic names do not indicate the structure of the compound. They are common for complex molecules, which includes most natural products. Thus, the informally named lysergic acid diethylamide is systematically named (6aR,9R)-N,N-diethyl-7-methyl-4,6,6a,7,8,9-hexahydroindolo-[4,3-fg] quinoline-9-carboxamide.

With the increased use of computing, other naming methods have evolved that are intended to be interpreted by machines. Two popular formats are SMILES and InChI.

Structural Drawings

Organic molecules are described more commonly by drawings or structural formulas, combinations of drawings and chemical symbols. The line-angle formula is simple and unambiguous. In this system, the endpoints and intersections of each line represent one carbon, and hydrogen atoms can either be notated explicitly or assumed to be present as implied by tetravalent carbon. The depiction of organic compounds with drawings is greatly simplified by the fact that carbon in almost all organic compounds has four bonds, nitrogen three, oxygen two, and hydrogen one.

Classification of Organic Compounds

Functional Groups

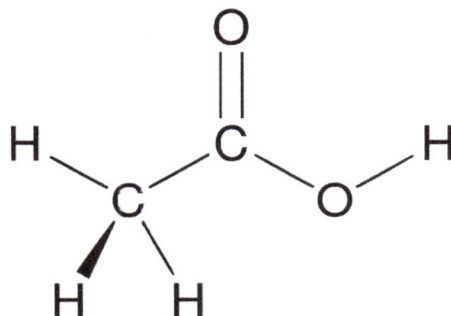

The family of carboxylic acids contains a carboxyl (-COOH) functional group. Acetic acid, shown here, is an example.

The concept of functional groups is central in organic chemistry, both as a means to classify structures and for predicting properties. A functional group is a molecular module, and the reactivity of that functional group is assumed, within limits, to be the same in a variety of molecules. Functional groups can have decisive influence on the chemical and physical properties of organic compounds. Molecules are classified on the basis of their functional groups. Alcohols, for example, all have the subunit C-O-H. All alcohols tend to be somewhat hydrophilic, usually form esters, and usually can be converted to the corresponding halides. Most functional groups feature heteroatoms (atoms other than C and H). Organic compounds are classified according to functional groups, alcohols, carboxylic acids, amines, etc.

Aliphatic Compounds

The aliphatic hydrocarbons are subdivided into three groups of homologous series according to their state of saturation:

- paraffins, which are alkanes without any double or triple bonds,

- olefins or alkenes which contain one or more double bonds, i.e. di-olefins (dienes) or poly-olefins.

- alkynes, which have one or more triple bonds.

The rest of the group is classed according to the functional groups present. Such compounds can be "straight-chain", branched-chain or cyclic. The degree of branching affects characteristics, such as the octane number or cetane number in petroleum chemistry.

Both saturated (alicyclic) compounds and unsaturated compounds exist as cyclic derivatives. The most stable rings contain five or six carbon atoms, but large rings (macrocycles) and smaller rings are common. The smallest cycloalkane family is the three-membered cyclopropane ($(CH_2)_3$). Saturated cyclic compounds contain single bonds only, whereas aromatic rings have an alternating (or conjugated) double bond. Cycloalkanes do not contain multiple bonds, whereas the cycloalkenes and the cycloalkynes do.

Aromatic Compounds

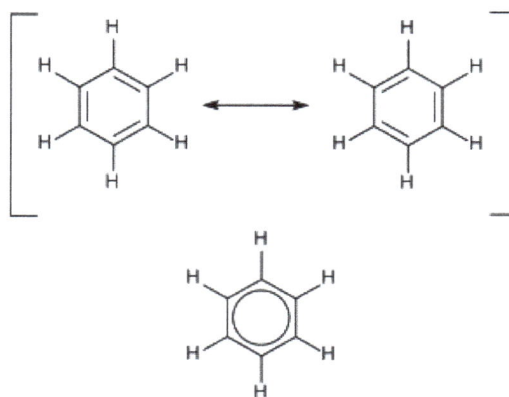

Benzene is one of the best-known aromatic compounds as it is one of the simplest and most stable aromatics.

Aromatic hydrocarbons contain conjugated double bonds. This means that every carbon atom in the ring is sp2 hybridized, allowing for added stability. The most important example is benzene, the structure of which was formulated by Kekulé who first proposed the delocalization or resonance principle for explaining its structure. For "conventional" cyclic compounds, aromaticity is conferred by the presence of 4n + 2 delocalized pi electrons, where n is an integer. Particular instability (antiaromaticity) is conferred by the presence of 4n conjugated pi electrons.

Heterocyclic Compounds

The characteristics of the cyclic hydrocarbons are again altered if heteroatoms are present, which can exist as either substituents attached externally to the ring (exocyclic) or as a member of the ring itself (endocyclic). In the case of the latter, the ring is termed a heterocycle. Pyridine and furan are examples of aromatic heterocycles while piperidine and tetrahydrofuran are the corresponding alicyclic heterocycles. The heteroatom of heterocyclic molecules is generally oxygen, sulfur, or nitrogen, with the latter being particularly common in biochemical systems.

Heterocycles are commonly found in a wide range of products including aniline dyes and medicines. Additionally, they are prevalent in a wide range of biochemical compounds such as alkaloids, vitamins, steroids, and nucleic acids (e.g. DNA, RNA).

Rings can fuse with other rings on an edge to give polycyclic compounds. The purine nucleoside bases are notable polycyclic aromatic heterocycles. Rings can also fuse on a "corner" such that one atom (almost always carbon) has two bonds going to one ring and two to another. Such compounds are termed spiro and are important in a number of natural products.

Polymers

This swimming board is made of polystyrene, an example of a polymer.

One important property of carbon is that it readily forms chains, or networks, that are linked by carbon-carbon (carbon-to-carbon) bonds. The linking process is called polymerization, while the chains, or networks, are called polymers. The source compound is called a monomer.

Two main groups of polymers exist: synthetic polymers and biopolymers. Synthetic polymers are artificially manufactured, and are commonly referred to as industrial polymers. Biopolymers occur within a respectfully natural environment, or without human intervention.

Since the invention of the first synthetic polymer product, bakelite, synthetic polymer products have frequently been invented.

Common synthetic organic polymers are polyethylene (polythene), polypropylene, nylon, teflon (PTFE), polystyrene, polyesters, polymethylmethacrylate (called perspex and plexiglas), and polyvinylchloride (PVC).

Both synthetic and natural rubber are polymers.

Varieties of each synthetic polymer product may exist, for purposes of a specific use. Changing the conditions of polymerization alters the chemical composition of the product and its properties. These alterations include the chain length, or branching, or the tacticity.

With a single monomer as a start, the product is a homopolymer.

Secondary component(s) may be added to create a heteropolymer (co-polymer) and the degree of clustering of the different components can also be controlled.

Physical characteristics, such as hardness, density, mechanical or tensile strength, abrasion resistance, heat resistance, transparency, colour, etc. will depend on the final composition.

Biomolecules

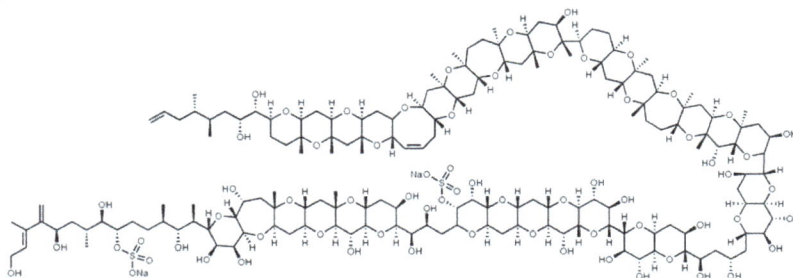

Maitotoxin, a complex organic biological toxin.

Biomolecular chemistry is a major category within organic chemistry which is frequently studied by biochemists. Many complex multi-functional group molecules are important in living organisms. Some are long-chain biopolymers, and these include peptides, DNA, RNA and the polysaccharides such as starches in animals and celluloses in plants. The other main classes are amino acids (monomer building blocks of peptides and proteins), carbohydrates (which includes the polysaccharides), the nucleic acids (which include DNA and RNA as polymers), and the lipids. In addition, animal biochemistry contains many small molecule intermediates which assist in energy production through the Krebs cycle, and produces isoprene, the most common hydrocarbon in animals. Isoprenes in animals form the important steroid structural (cholesterol) and steroid hormone compounds; and in plants form terpenes, terpenoids, some alkaloids, and a class of hydrocarbons called biopolymer polyisoprenoids present in the latex of various species of plants, which is the basis for making rubber.

Small Molecules

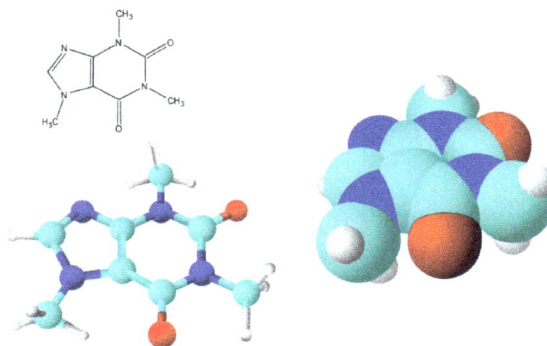

Molecular models of caffeine.

In pharmacology, an important group of organic compounds is small molecules, also referred to as 'small organic compounds'. In this context, a small molecule is a small organic compound that is biologically active, but is not a polymer. In practice, small molecules have a molar mass less than approximately 1000 g/mol.

Fullerenes

Fullerenes and carbon nanotubes, carbon compounds with spheroidal and tubular structures, have stimulated much research into the related field of materials science. The first fullerene was discovered in 1985 by Sir Harold W. Kroto (one of the authors of this article) of the United Kingdom and by Richard E. Smalley and Robert F. Curl, Jr., of the United States. Using a laser to vaporize graphite rods in an atmosphere of helium gas, these chemists and their assistants obtained cagelike molecules composed of 60 carbon atoms (C60) joined together by single and double bonds to form a hollow sphere with 12 pentagonal and 20 hexagonal faces—a design that resembles a football, or soccer ball. In 1996 the trio was awarded the Nobel Prize for their pioneering efforts. The C60 molecule was named buckminsterfullerene (or, more simply, the buckyball) after the American architect R. Buckminster Fuller, whose geodesic dome is constructed on the same structural principles.

Others

Organic compounds containing bonds of carbon to nitrogen, oxygen and the halogens are not normally grouped separately. Others are sometimes put into major groups within organic chemistry and discussed under titles such as organosulfur chemistry, organometallic chemistry, organophosphorus chemistry and organosilicon chemistry.

Organic Reactions

Organic reactions are chemical reactions involving organic compounds. Many of these reactions are associated with functional groups. The general theory of these reactions involves careful analysis of such properties as the electron affinity of key atoms, bond strengths and steric hindrance. These factors can determine the relative stability of short-lived reactive intermediates, which usually directly determine the path of the reaction.

The basic reaction types are: addition reactions, elimination reactions, substitution reactions, pericyclic reactions, rearrangement reactions and redox reactions. An example of a common reaction is a substitution reaction written as:

$$Nu^- + C\text{-}X \rightarrow C\text{-}Nu + X^-$$

where X is some functional group and Nu is a nucleophile.

The number of possible organic reactions is basically infinite. However, certain general patterns are observed that can be used to describe many common or useful reactions. Each reaction has a stepwise reaction mechanism that explains how it happens in sequence—although the detailed description of steps is not always clear from a list of reactants alone.

The stepwise course of any given reaction mechanism can be represented using arrow pushing techniques in which curved arrows are used to track the movement of electrons as starting materials transition through intermediates to final products.

Organic Synthesis

Synthetic organic chemistry is an applied science as it borders engineering, the "design, analysis, and/or construction of works for practical purposes". Organic synthesis of a novel compound is a problem solving task, where a synthesis is designed for a target molecule by selecting optimal reactions from optimal starting materials. Complex compounds can have tens of reaction steps that sequentially build the desired molecule. The synthesis proceeds by utilizing the reactivity of the functional groups in the molecule. For example, a carbonyl compound can be used as a nucleophile by converting it into an enolate, or as an electrophile; the combination of the two is called the aldol reaction. Designing practically useful syntheses always requires conducting the actual synthesis in the laboratory. The scientific practice of creating novel synthetic routes for complex molecules is called total synthesis.

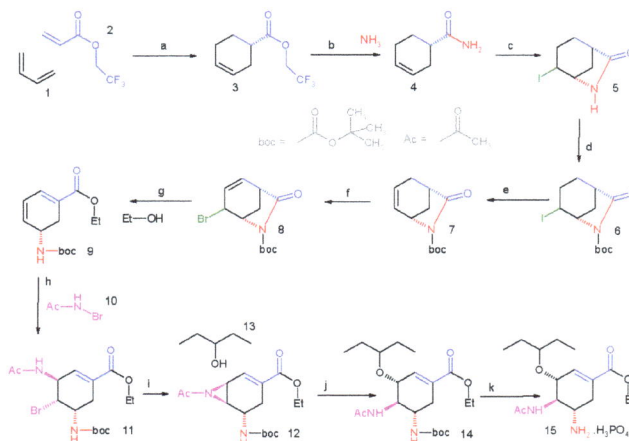

A synthesis designed by E.J. Corey for oseltamivir (Tamiflu). This synthesis has 11 distinct reactions.

Strategies to design a synthesis include retrosynthesis, popularized by E.J. Corey, starts with the target molecule and splices it to pieces according to known reactions. The pieces, or the proposed precursors, receive the same treatment, until available and ideally inexpensive starting materials are reached. Then, the retrosynthesis is written in the opposite direction to give the synthesis. A "synthetic tree" can be constructed, because each compound and also each precursor has multiple syntheses.

References

- Assuring a Future U.S.-Based Nuclear and Radiochemistry Expertise. Board on Chemical Sciences and Technology. 2012. ISBN 978-0-309-22534-2.

- Torben Smith Sørensen (1999). Surface chemistry and electrochemistry of membranes. CRC Press. p. 134. ISBN 0-8247-1922-0.

- Atkins, Peter and Friedman, Ronald (2005). Molecular Quantum Mechanics, p. 249. Oxford University Press, New York. ISBN 0-19-927498-3.

- Landau, L. D. and Lifshitz, E. M. (1980). Statistical Physics, 3rd Ed. p. 52. Elsevier Butterworth Heinemann, New York. ISBN 0-7506-3372-7.

- Hill, Terrell L. (1986). Introduction to Statistical Thermodynamics, p. 1. Dover Publications, New York. ISBN 0-486-65242-4.

- Schmidt, Lanny D. (2005). The Engineering of Chemical Reactions, 2nd Ed. p. 30. Oxford University Press, New York. ISBN 0-19-516925-5.

- Chandler, David (1987). Introduction to Modern Statistical Mechanics, p. 54. Oxford University Press, New York. ISBN 978-0-19-504277-1.

- Greenwood, Norman N.; Earnshaw, Alan (1997). Chemistry of the Elements (2nd ed.). Butterworth-Heinemann. ISBN 0-08-037941-9.

- Nicolaou, K. C.; Sorensen, E. J. (1996). Classics in Total Synthesis: Targets, Strategies, Methods. Wiley. ISBN 978-3-527-29231-8.

- Allan, Barbara. Livesey, Brian (1994). How to Use Biological Abstracts, Chemical Abstracts and Index Chemicus. Gower. ISBN 978-0566075568

- Shriner, R.L.; Hermann, C.K.F.; Morrill, T.C.; Curtin, D.Y. and Fuson, R.C. (1997) The Systematic Identification of Organic Compounds. John Wiley & Sons, ISBN 0-471-59748-1

- Jordans, Frank (July 30, 2015). "Philae probe finds evidence that comets can be cosmic labs". The Washington Post. Associated Press. Retrieved July 30, 2015.

- Bibring, J.-P.; Taylor, M.G.G.T.; Alexander, C.; Auster, U.; Biele, J.; Finzi, A. Ercoli; Goesmann, F.; Klingehoefer, G.; Kofman, W.; Mottola, S.; Seidenstiker, K.J.; Spohn, T.; Wright, I. (July 31, 2015).

- "What Is Organic Chemistry?". Science Alive! The Life and Science of Percy Julian. Chemical Heritage Foundation. Retrieved 27 January 2015.

- "NASA Ames PAH IR Spectroscopic Database". The Astrophysics & Astrochemistry Laboratory, NASA-Ames. 29 Oct 2013. Retrieved 18 Apr 2014.

- Hoover, Rachel (February 21, 2014). "Need to Track Organic Nano-Particles Across the Universe? NASA's Got an App for That". NASA. Retrieved February 22, 2014.

- Zubritsky, Elizabeth; Neal-Jones, Nancy (August 11, 2014). "RELEASE 14-038 - NASA's 3-D Study of Comets Reveals Chemical Factory at Work". NASA. Retrieved August 12, 2014.

- "Thermodynamic models of the distribution of life-related organic molecules in the interstellar medium". Astrophysics and Space Science. May 2014. Retrieved 25 Feb 2014.

- "About Organic Systems Origin According to Equilibrium Thermodynamic Modeles of Molecules Distribution in Interstellar Medium". Canadian Center of Science and Education. 20 July 2014. Retrieved 4 Aug 2014.

Quantum Chemistry: An Integrated Study

Quantum chemistry is a branch of chemistry that deals with the application of quantum mechanics in experiments and physical models of chemical systems. Quantum chemistry is an emerging field of study and the following chapter will not only provide an overview, but it will also inquire deep into the diverse topics related to it.

Quantum chemistry is a branch of chemistry whose primary focus is the application of quantum mechanics in physical models and experiments of chemical systems. It is also called molecular quantum mechanics.

Overview

It involves heavy interplay of experimental and theoretical methods:

- Experimental quantum chemists rely heavily on spectroscopy, through which information regarding the quantization of energy on a molecular scale can be obtained. Common methods are infra-red (IR) spectroscopy and nuclear magnetic resonance (NMR) spectroscopy.

- Theoretical quantum chemistry, the workings of which also tend to fall under the category of computational chemistry, seeks to calculate the predictions of quantum theory as atoms and molecules can only have discrete energies; as this task, when applied to polyatomic species, invokes the many-body problem, these calculations are performed using computers rather than by analytical "back of the envelope" methods, pen recorder or computerized data station with a VDU.

In these ways, quantum chemists investigate chemical phenomena.

- In reactions, quantum chemistry studies the ground state of individual atoms and molecules, the excited states, and the transition states that occur during chemical reactions.

- On the calculations: quantum chemical studies use also semi-empirical and other methods based on quantum mechanical principles, and deal with time dependent problems. Many quantum chemical studies assume the nuclei are at rest (Born–Oppenheimer approximation). Many calculations involve iterative methods that include self-consistent field methods. Major goals of quantum chemistry include increasing the accuracy of the results for small molecular systems, and increasing the size of large molecules that can be processed, which is limited by scaling considerations—the computation time increases as a power of the number of atoms.

History

Some view the birth of quantum chemistry in the discovery of the Schrödinger equation and its application to the hydrogen atom in 1926. However, the 1927 article of Walter Heitler and Fritz London is often recognised as the first milestone in the history of quantum chemistry. This is the first application of quantum mechanics to the diatomic hydrogen molecule, and thus to the phenomenon of the chemical bond. In the following years much progress was accomplished by Edward Teller, Robert S. Mulliken, Max Born, J. Robert Oppenheimer, Linus Pauling, Erich Hückel, Douglas Hartree, Vladimir Aleksandrovich Fock, to cite a few. The history of quantum chemistry also goes through the 1838 discovery of cathode rays by Michael Faraday, the 1859 statement of the black body radiation problem by Gustav Kirchhoff, the 1877 suggestion by Ludwig Boltzmann that the energy states of a physical system could be discrete, and the 1900 quantum hypothesis by Max Planck that any energy radiating atomic system can theoretically be divided into a number of discrete energy elements ε such that each of these energy elements is proportional to the frequency v with which they each individually radiate energy and a numerical value called Planck's Constant. Then, in 1905, to explain the photoelectric effect (1839), i.e., that shining light on certain materials can function to eject electrons from the material, Albert Einstein postulated, based on Planck's quantum hypothesis, that light itself consists of individual quantum particles, which later came to be called photons (1926). In the years to follow, this theoretical basis slowly began to be applied to chemical structure, reactivity, and bonding. Probably the greatest contribution to the field was made by Linus Pauling.

Electronic Structure

The first step in solving a quantum chemical problem is usually solving the Schrödinger equation (or Dirac equation in relativistic quantum chemistry) with the electronic molecular Hamiltonian. This is called determining the electronic structure of the molecule. It can be said that the electronic structure of a molecule or crystal implies essentially its chemical properties. An exact solution for the Schrödinger equation can only be obtained for the hydrogen atom (though exact solutions for the bound state energies of the hydrogen molecular ion have been identified in terms of the generalized Lambert W function). Since all other atomic, or molecular systems, involve the motions of three or more "particles", their Schrödinger equations cannot be solved exactly and so approximate solutions must be sought.

Wave Model

The foundation of quantum mechanics and quantum chemistry is the wave model, in which the atom is a small, dense, positively charged nucleus surrounded by electrons. The wave model is derived from the wavefunction, a set of possible equations derived from the time evolution of the Schrödinger equation which is applied to the wavelike probability distribution of subatomic particles. Unlike the earlier Bohr model of the atom, however, the wave model describes electrons as "clouds" moving in orbitals, and their positions are represented by probability distributions rather than discrete points. The strength of this model lies in its predictive power. Specifically, it predicts the pattern of chemically similar elements found in the periodic table. The wave model is so named because electrons exhibit properties (such as interference) traditionally associated with waves. See wave-particle duality. In this model, when we solve the Schrödinger Equation for an Hidrogenoid Atom, we obtain a solution that depends on some numbers, called quantum numbers, that describes the orbital, the most probable space where an electron can be. These are n, the principal quantum number, for the energy, l, or secondary quantum number, which correlates to the angular momentum, ml, for the orientation, and ms the spin. This model can explain the new lines that appeared in the spectroscopy of atoms. For multielectron atoms we must introduce some rules as that the electrons fill orbitals in a way to minimize the energy of the atom, in order of increasing energy, the Pauli Exclusion Principle, the Hund's Rule, and the Aufbau Principle.

Valence Bond

Although the mathematical basis of quantum chemistry had been laid by Schrödinger in 1926, it is generally accepted that the first true calculation in quantum chemistry was that of the German physicists Walter Heitler and Fritz London on the hydrogen (H_2) molecule in 1927. Heitler and London's method was extended by the American theoretical physicist John C. Slater and the American theoretical chemist Linus Pauling to become the Valence-Bond (VB) [or Heitler–London–Slater–Pauling (HLSP)] method. In this method, attention is primarily devoted to the pairwise interactions between atoms, and this method therefore correlates closely with classical chemists' drawings of bonds. It focuses on how the atomic orbitals of an atom combine to give individual chemical bonds when a molecule is formed.

Molecular Orbital

An alternative approach was developed in 1929 by Friedrich Hund and Robert S. Mulliken, in which electrons are described by mathematical functions delocalized over an entire molecule. The Hund–Mulliken approach or molecular orbital (MO) method is less intuitive to chemists, but has turned out capable of predicting spectroscopic properties better than the VB method. This approach is the conceptional basis of the Hartree–Fock method and further post Hartree–Fock methods.

Density Functional Theory

The Thomas–Fermi model was developed independently by Thomas and Fermi in 1927. This was the first attempt to describe many-electron systems on the basis of electronic density instead of wave functions, although it was not very successful in the treatment of entire molecules. The method did provide the basis for what is now known as density functional theory. Modern day DFT uses the Kohn-Sham method, where the density functional is split into four terms; the Kohn-Sham kinetic energy, an external potential, exchange and correlation energies. A large part of the focus on developing DFT is on improving the exchange and correlation terms. Though this method is less developed than post Hartree–Fock methods, its significantly lower computational requirements (scaling typically no worse than with respect to basis functions, for the pure functionals) allow it to tackle larger polyatomic molecules and even macromolecules. This computational affordability and often comparable accuracy to MP2 and CCSD(T) (post-Hartree–Fock methods) has made it one of the most popular methods in computational chemistry at present.

Chemical Dynamics

A further step can consist of solving the Schrödinger equation with the total molecular Hamiltonian in order to study the motion of molecules. Direct solution of the Schrödinger equation is called *quantum molecular dynamics*, within the semiclassical approximation *semiclassical molecular dynamics*, and within the classical mechanics framework *molecular dynamics (MD)*. Statistical approaches, using for example Monte Carlo methods, are also possible.

Adiabatic Chemical Dynamics

In adiabatic dynamics, interatomic interactions are represented by single scalar potentials called potential energy surfaces. This is the Born–Oppenheimer approximation introduced by Born and Oppenheimer in 1927. Pioneering applications of this in chemistry were performed by Rice and Ramsperger in 1927 and Kassel in 1928, and generalized into the RRKM theory in 1952 by Marcus who took the transition state theory developed by Eyring in 1935 into account. These methods enable simple estimates of unimolecular reaction rates from a few characteristics of the potential surface.

Non-Adiabatic Chemical Dynamics

Non-adiabatic dynamics consists of taking the interaction between several coupled potential energy surface (corresponding to different electronic quantum states of the molecule). The coupling terms are called vibronic couplings. The pioneering work in this field was done by Stueckelberg, Landau, and Zener in the 1930s, in their work on what is now known as the Landau–Zener transition. Their formula allows the transition probability between two diabatic potential curves in the neighborhood of an avoided crossing to be calculated.

References

- Atkins, P.W.; Friedman, R. (2005). Molecular Quantum Mechanics (4th ed.). Oxford University Press. ISBN 978-0-19-927498-7.

- Pullman, Bernard; Pullman, Alberte (1963). Quantum Biochemistry. New York and London: Academic Press. ISBN 90-277-1830-X.

- Kostas Gavroglu, Ana Simões: NEITHER PHYSICS NOR CHEMISTRY.A History of Quantum Chemistry, MIT Press, 2011, ISBN 0-262-01618-4

- Karplus M., Porter R.N. (1971). Atoms and Molecules. An introduction for students of physical chemistry, Benjamin–Cummings Publishing Company, ISBN 978-0-8053-5218-4

- Szabo, Attila; Ostlund, Neil S. (1996). Modern Quantum Chemistry: Introduction to Advanced Electronic Structure Theory. Dover. ISBN 0-486-69186-1.

- Landau, L.D.; Lifshitz, E.M. Quantum Mechanics:Non-relativistic Theory. Course of Theoretical Physic 3. Pergamon Press. ISBN 0-08-019012-X.

- Pauling, L.; Wilson, E. B. (1963) . Introduction to Quantum Mechanics with Applications to Chemistry. Dover Publications. ISBN 0-486-64871-0.

Diverse Chemical Compounds, Bonds and Reactions

This chapter elucidates the crucial theories and principles on chemical compounds. A chemical compound is an entity consisting of two or more atoms; there are four types of compounds, depending on how the constituents are held together. The topics discussed are of great importance to broaden the existing knowledge on chemistry.

Chemical Bonds

A chemical bond is a lasting attraction between atoms that enables the formation of chemical compounds. The bond may result from the electrostatic force of attraction between atoms with opposite charges, or through the sharing of electrons as in the covalent bonds. The strength of chemical bonds varies considerably; there are "strong bonds" such as covalent or ionic bonds and "weak bonds" such as Dipole-dipole interaction, the London dispersion force and hydrogen bonding.

Since opposite charges attract via a simple electromagnetic force, the negatively charged electrons that are orbiting the nucleus and the positively charged protons in the nucleus attract each other. An electron positioned between two nuclei will be attracted to both of them, and the nuclei will be attracted toward electrons in this position. This attraction constitutes the chemical bond. Due to the matter wave nature of electrons and their smaller mass, they must occupy a much larger amount of volume compared with the nuclei, and this volume occupied by the electrons keeps the atomic nuclei relatively far apart, as compared with the size of the nuclei themselves. This phenomenon limits the distance between nuclei and atoms in a bond.

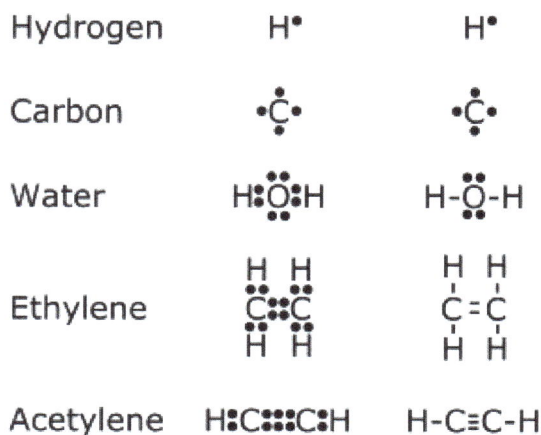

Examples of Lewis dot-style representations of chemical bonds between carbon (C), hydrogen (H), and oxygen (O). Lewis dot diagrams were an early attempt to describe chemical bonding and are still widely used today.

In general, strong chemical bonding is associated with the sharing or transfer of electrons between the participating atoms. The atoms in molecules, crystals, metals and diatomic gases—indeed most of the physical environment around us—are held together by chemical bonds, which dictate the structure and the bulk properties of matter.

All bonds can be explained by quantum theory, but, in practice, simplification rules allow chemists to predict the strength, directionality, and polarity of bonds. The octet rule and VSEPR theory are two examples. More sophisticated theories are valence bond theory which includes orbital hybridization and resonance, and molecular orbital theory which includes linear combination of atomic orbitals and ligand field theory. Electrostatics are used to describe bond polarities and the effects they have on chemical substances.

Overview of Main Types of Chemical Bonds

A chemical bond is an attraction between atoms. This attraction may be seen as the result of different behaviors of the outermost or valence electrons of atoms. Although all of these behaviors merge into each other seamlessly in various bonding situations so that there is no clear line to be drawn between them, the behaviors of atoms become so qualitatively different as the character of the bond changes quantitatively, that it remains useful and customary to differentiate between the bonds that cause these different properties of condensed matter.

In the simplest view of a covalent bond one or more electrons (often a pair of electrons) are drawn into the space between the two atomic nuclei. The reduction in energy from bond formation comes not from the reduction in potential energy, because the attraction of the two electrons to the two protons is offset by the electron-electron and proton-proton repulsions. Instead, the reduction in overall energy (and hence stability of the bond) arises from the reduction in kinetic energy due to the electrons being in a more spatially distributed (i.e., longer deBroglie wavelength) orbital compared with each electron being confined closer to its respective nucleus. These bonds exist between two particular identifiable atoms and have a direction in space, allowing them to be shown as single connecting lines between atoms in drawings, or modeled as sticks between spheres in models.

In a polar covalent bond, one or more electrons are unequally shared between two nuclei. Covalent bonds often result in the formation of small collections of better-connected atoms called molecules, which in solids and liquids are bound to other molecules by forces that are often much weaker than the covalent bonds that hold the molecules internally together. Such weak intermolecular bonds give organic molecular substances, such as waxes and oils, their soft bulk character, and their low melting points (in liquids, molecules must cease most structured or oriented contact with each other). When covalent bonds link long chains of atoms in large molecules, however (as in polymers such as nylon), or when covalent bonds extend in networks through solids that are not composed of discrete molecules (such as diamond or quartz or the silicate minerals in many types of rock) then the structures that result may be both strong and tough, at least in the direction oriented correctly with networks of covalent bonds. Also, the melting points of such covalent polymers and networks increase greatly.

In a simplified view of an *ionic* bond, the bonding electron is not shared at all, but transferred. In this type of bond, the outer atomic orbital of one atom has a vacancy which allows the addition of one or more electrons. These newly added electrons potentially occupy a lower energy-state (effectively closer to more nuclear charge) than they experience in a different atom. Thus, one nucleus offers a more tightly bound position to an electron than does another nucleus, with the result that one atom may transfer an electron to the other. This transfer causes one atom to assume a net positive charge, and the other to assume a net negative charge. The *bond* then results from electrostatic attraction between atoms and the atoms become positive or negatively charged ions. Ionic bonds may be seen as extreme examples of polarization in covalent bonds. Often, such bonds have no particular orientation in space, since they result from equal electrostatic attraction of each ion to all ions around them. Ionic bonds are strong (and thus ionic substances require high temperatures to melt) but also brittle, since the forces between ions are short-range and do not easily bridge cracks and fractures. This type of bond gives rise to the physical characteristics of crystals of classic mineral salts, such as table salt.

A less often mentioned type of bonding is *metallic* bonding. In this type of bonding, each atom in a metal donates one or more electrons to a "sea" of electrons that reside between many metal atoms. In this sea, each electron is free (by virtue of its wave nature) to be associated with a great many atoms at once. The bond results because the metal atoms become somewhat positively charged due to loss of their electrons while the electrons remain attracted to many atoms, without being part of any given atom. Metallic bonding may be seen as an extreme example of delocalization of electrons over a large system of covalent bonds, in which every atom participates. This type of bonding is often very strong (resulting in the tensile strength of metals). However, metallic bonding is more collective in nature than other types, and so they allow metal crystals to more easily deform, because they are composed of atoms attracted to each other, but not in any particularly-oriented ways. This results in the malleability of metals. The sea of electrons in metallic bonding causes the characteristically good electrical and thermal conductivity of metals, and also their "shiny" reflection of most frequencies of white light.

History

Early speculations into the nature of the chemical bond, from as early as the 12th century, supposed that certain types of chemical species were joined by a type of chemical affinity. In 1704, Sir Isaac Newton famously outlined his atomic bonding theory, in "Query 31" of his Opticks, whereby atoms attach to each other by some "force". Specifically, after acknowledging the various popular theories in vogue at the time, of how atoms were reasoned to attach to each other, i.e. "hooked atoms", "glued together by rest", or "stuck together by conspiring motions", Newton states that he would rather infer from their cohesion, that "particles attract one another by some force, which in immediate contact is exceedingly strong, at small distances performs the chemical operations, and reaches not far from the particles with any sensible effect."

In 1819, on the heels of the invention of the voltaic pile, Jöns Jakob Berzelius developed a theory of chemical combination stressing the electronegative and electropositive character of the combining atoms. By the mid 19th century, Edward Frankland, F.A. Kekulé, A.S. Couper, Alexander Butlerov, and Hermann Kolbe, building on the theory of radicals, developed the theory of valency, originally called "combining power", in which compounds were joined owing to an attraction of positive and negative poles. In 1916, chemist Gilbert N. Lewis developed the concept of the electron-pair bond, in which two atoms may share one to six electrons, thus forming the single electron bond, a single bond, a double bond, or a triple bond; in Lewis's own words, "An electron may form a part of the shell of two different atoms and cannot be said to belong to either one exclusively."

That same year, Walther Kossel put forward a theory similar to Lewis' only his model assumed complete transfers of electrons between atoms, and was thus a model of ionic bonding. Both Lewis and Kossel structured their bonding models on that of Abegg's rule (1904).

In 1927, the first mathematically complete quantum description of a simple chemical bond, i.e. that produced by one electron in the hydrogen molecular ion, H_2^+, was derived by the Danish physicist Oyvind Burrau. This work showed that the quantum approach to chemical bonds could be fundamentally and quantitatively correct, but the mathematical methods used could not be extended to molecules containing more than one electron. A more practical, albeit less quantitative, approach was put forward in the same year by Walter Heitler and Fritz London. The Heitler-London method forms the basis of what is now called valence bond theory. In 1929, the linear combination of atomic orbitals molecular orbital method (LCAO) approximation was introduced by Sir John Lennard-Jones, who also suggested methods to derive electronic structures of molecules of F_2 (fluorine) and O_2 (oxygen) molecules, from basic quantum principles. This molecular orbital theory represented a covalent bond as an orbital formed by combining the quantum mechanical Schrödinger atomic orbitals which had been hypothesized for electrons in single atoms. The equations for bonding electrons in multi-electron atoms could not be solved to mathematical perfection (i.e., *analytically*), but approximations for them still gave many good qualitative predictions and results. Most quantitative calculations in modern quantum chemistry use either valence bond or molecular orbital theory as a starting point, although a third approach, density functional theory, has become increasingly popular in recent years.

In 1933, H. H. James and A. S. Coolidge carried out a calculation on the dihydrogen molecule that, unlike all previous calculation which used functions only of the distance of the electron from the atomic nucleus, used functions which also explicitly added the distance between the two electrons. With up to 13 adjustable parameters they obtained a result very close to the experimental result for the dissociation energy. Later extensions have used up to 54 parameters and gave excellent agreement with experiments. This calculation convinced the scientific community that quantum theory could give agreement with experiment. However this approach has none of the physical pictures of the valence bond and molecular orbital theories and is difficult to extend to larger molecules.

Bonds in Chemical Formulas

The fact that atoms and molecules are three-dimensional makes it difficult to use a single technique for indicating orbitals and bonds. In molecular formulas the chemical bonds (binding orbitals) between atoms are indicated by various methods according to the type of discussion. Sometimes, they are completely neglected. For example, in organic chemistry chemists are sometimes concerned only with the functional group of the molecule. Thus, the molecular formula of ethanol may be written in a paper in conformational form, three-dimensional form, full two-dimensional form (indicating every bond with no three-dimensional directions), compressed two-dimensional form (CH_3-CH_2-OH), by separating the functional group from another part of the molecule (C_2H_5OH), or by its atomic constituents (C_2H_6O), according to what is discussed. Sometimes, even the non-bonding valence shell electrons (with the two-dimensional approximate directions) are marked, i.e. for elemental carbon $\dot{C}$. Some chemists may also mark the respective orbitals, i.e. the hypothetical ethene^{-4} anion ($_\backslash^/C=C_/^\backslash$ $^{-4}$) indicating the possibility of bond formation.

Strong Chemical Bonds

Typical bond lengths in pm and bond energies in kJ/mol. Bond lengths can be converted to Å by division by 100 (1 Å = 100 pm). Data taken from University of Waterloo.		
Bond	**Length (pm)**	**Energy (kJ/mol)**
H — Hydrogen		
H–H	74	436
H–O	96	366
H–F	92	568
H–Cl	127	432
C — Carbon		
C–H	109	413
C–C	154	348
C–C=	151	
=C–C⬜	147	
=C–C=	148	
C=C	134	614
C⬜C	120	839
C–N	147	308
C–O	143	360
C–F	134	488
C–Cl	177	330
N — Nitrogen		
N–H	101	391
N–N	145	170
N⬜N	110	945
O — Oxygen		
O–O	148	145
O=O	121	498
F, Cl, Br, I — Halogens		
F–F	142	158
Cl–Cl	199	243
Br–H	141	366
Br–Br	228	193
I–H	161	298
I–I	267	151

Strong chemical bonds are the *intramolecular* forces which hold atoms together in molecules. A strong chemical bond is formed from the transfer or sharing of electrons between atomic centers and relies on the electrostatic attraction between the protons in nuclei and the electrons in the orbitals.

The types of strong bond differ due to the difference in electronegativity of the constituent elements. A large difference in electronegativity leads to more polar (ionic) character in the bond.

Ionic Bonding

Ionic bonding is a type of electrostatic interaction between atoms which have a large electronegativity difference. There is no precise value that distinguishes ionic from covalent bonding, but a difference of electronegativity of over 1.7 is likely to be ionic, and a difference of less than 1.7 is likely to be covalent. Ionic bonding leads to separate positive and negative ions. Ionic charges are commonly between −3e to +3e. Ionic bonding commonly occurs in metal salts such as sodium chloride (table salt). A typical feature of ionic bonds is that the species form into ionic crystals, in which no ion is specifically paired with any single other ion, in a specific directional bond. Rather, each species of ion is surrounded by ions of the opposite charge, and the spacing between it and each of the oppositely charged ions near it, is the same for all surrounding atoms of the same type. It is thus no longer possible to associate an ion with any specific other single ionized atom near it. This is a situation unlike that in covalent crystals, where covalent bonds between specific atoms are still discernible from the shorter distances between them, as measured via such techniques as X-ray diffraction.

Ionic crystals may contain a mixture of covalent and ionic species, as for example salts of complex acids, such as sodium cyanide, NaCN. X-ray diffraction shows that in NaCN, for example, the bonds between sodium cations (Na^+) and the cyanide anions (CN^-) are *ionic*, with no sodium ion associated with any particular cyanide. However, the bonds between C and N atoms in cyanide are of the *covalent* type, making each of the carbon and nitrogen associated with *just one* of its opposite type, to which it is physically much closer than it is to other carbons or nitrogens in a sodium cyanide crystal.

When such crystals are melted into liquids, the ionic bonds are broken first because they are non-directional and allow the charged species to move freely. Similarly, when such salts dissolve into water, the ionic bonds are typically broken by the interaction with water, but the covalent bonds continue to hold. For example, in solution, the cyanide ions, still bound together as single CN^- ions, move independently through the solution, as do sodium ions, as Na^+. In water, charged ions move apart because each of them are more strongly attracted to a number of water molecules, than to each other. The attraction between ions and water molecules in such solutions is due to a type of weak dipole-dipole type chemical bond. In melted ionic compounds, the ions continue to be attracted to each other, but not in any ordered or crystalline way.

Covalent Bond

Covalent bonding is a common type of bonding, in which two atoms share two valence electrons, one from each of the atoms. In nonpolar covalent bonds, the electronegativity difference between the bonded atoms is small, typically 0 to 0.3. Bonds within most organic compounds are described as covalent. The figure shows methane (CH_4), in which each hydrogen forms a covalent bond with the carbon. See sigma bonds and pi bonds for LCAO-description of such bonding.

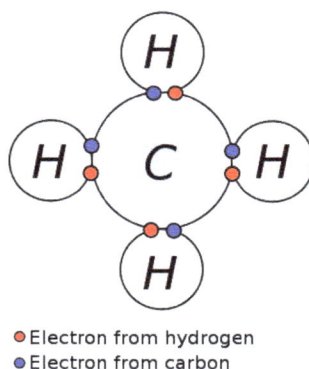

● Electron from hydrogen
● Electron from carbon

Nonpolar covalent bonds in methane (CH_4). The Lewis structure shows electrons shared between C and H atoms.

Molecules which are formed primarily from non-polar covalent bonds are often immiscible in water or other polar solvents, but much more soluble in non-polar solvents such as hexane.

A polar covalent bond is a covalent bond with a significant ionic character. This means that the two shared electrons are closer to one of the atoms than the other, creating an imbalance of charge. Such bonds occur between two atoms with moderately different electronegativities and give rise to dipole-dipole interactions. The electronegativity difference between the two atoms in these bonds is 0.3 to 1.7.

Single and Multiple Bonds

A single bond between two atoms corresponds to the sharing of one pair of electrons. The electron density of these two bonding electrons is concentrated in the region between the two atoms, which is the defining quality of a sigma bond.

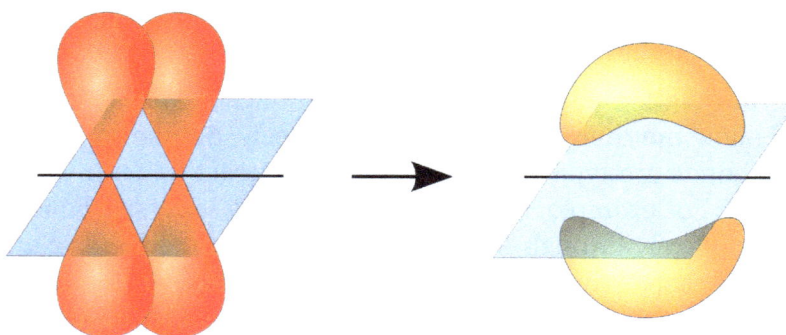

Two p-orbitals forming a pi-bond.

A double bond between two atoms is formed by the sharing of two pairs of electrons, one in a sigma bond and one in a pi bond, with electron density concentrated on two opposite sides of the internuclear axis. A triple bond consists of three shared electron pairs, forming one sigma and two pi bonds.

Quadruple and higher bonds are very rare and occur only between certain transition metal atoms.

Coordinate Covalent Bond (Dipolar Bond)

Adduct of ammonia and boron trifluoride

A coordinate covalent bond is a covalent bond in which the two shared bonding electrons are from the same one of the atoms involved in the bond. For example, boron trifluoride (BF_3) and ammonia (NH_3) from an adduct or coordination complex $F_3B\leftarrow NH_3$ with a B–N bond in which a lone pair of electrons on N is shared with an empty atomic orbital on B. BF_3 with an empty orbital is described as an electron pair acceptor or Lewis acid, while NH_3 with a lone pair which can be shared is described as an electron-pair donor or Lewis base. The electrons are shared roughly equally between the atoms in contrast to ionic bonding. Such bonding is shown by an arrow pointing to the Lewis acid.

Transition metal complexes are generally bound by coordinate covalent bonds. For example the ion Ag^+ reacts as a Lewis acid with two molecules of the Lewis base NH_3 to form the complex ion $Ag(NH_3)_2^+$, which has two $Ag\leftarrow N$ coordinate covalent bonds.

Metallic Bonding

In metallic bonding, bonding electrons are delocalized over a lattice of atoms. By contrast, in ionic compounds, the locations of the binding electrons and their charges are static. The freely-moving or delocalization of bonding electrons leads to classical metallic properties such as luster (surface light reflectivity), electrical and thermal conductivity, ductility, and high tensile strength.

Intermolecular Bonding

There are four basic types of bonds that can be formed between two or more (otherwise non-associated) molecules, ions or atoms. Intermolecular forces cause molecules to be attracted or repulsed by each other. Often, these define some of the physical characteristics (such as the melting point) of a substance.

- A large difference in electronegativity between two bonded atoms will cause a permanent charge separation, or dipole, in a molecule or ion. Two or more molecules or ions with permanent dipoles can interact within dipole-dipole interactions. The bonding electrons in a molecule or ion will, on average, be closer to the more electronegative atom more frequently than the less electronegative one, giving rise to partial charges on each atom, and causing electrostatic forces between molecules or ions.

- A hydrogen bond is effectively a strong example of an interaction between two permanent dipoles. The large difference in electronegativities between hydrogen and any of fluorine, nitrogen and oxygen, coupled with their lone pairs of electrons cause strong electrostatic forces between molecules. Hydrogen bonds are responsible for the high boiling points of water and ammonia with respect to their heavier analogues.

- The London dispersion force arises due to instantaneous dipoles in neighbouring atoms. As the negative charge of the electron is not uniform around the whole atom, there is always a charge imbalance. This small charge will induce a corresponding dipole in a nearby molecule; causing an attraction between the two. The electron then moves to another part of the electron cloud and the attraction is broken.

- A cation–pi interaction occurs between a pi bond and a cation.

Theories of Chemical Bonding

In the (unrealistic) limit of "pure" ionic bonding, electrons are perfectly localized on one of the two atoms in the bond. Such bonds can be understood by classical physics. The forces between the atoms are characterized by isotropic continuum electrostatic potentials. Their magnitude is in simple proportion to the charge difference.

Covalent bonds are better understood by valence bond theory or molecular orbital theory. The properties of the atoms involved can be understood using concepts such as oxidation number. The electron density within a bond is not assigned to individual atoms, but is instead delocalized between atoms. In valence bond theory, the two electrons on the two atoms are coupled together with the bond strength depending on the overlap between them. In molecular orbital theory, the linear combination of atomic orbitals (LCAO) helps describe the delocalized molecular orbital structures and energies based on the atomic orbitals of the atoms they came from. Unlike pure ionic bonds, covalent bonds may have directed anisotropic properties. These may have their own names, such as sigma bond and pi bond.

In the general case, atoms form bonds that are intermediate between ionic and covalent, depending on the relative electronegativity of the atoms involved. This type of bond is sometimes called polar covalent.

Chemical Compound

A chemical compound (or just compound if used in the context of chemistry) is an entity consisting of two or more atoms, at least two from different elements, which associate via chemical bonds. There are four types of compounds, depending on how the constituent atoms are held together: molecules held together by covalent bonds, salts held together by ionic bonds, intermetallic compounds held together by metallic bonds, and certain complexes held together by coordinate covalent bonds. Many chemical compounds have a unique numerical identifier assigned by the Chemical Abstracts Service (CAS): its CAS number.

A chemical formula is a way of expressing information about the proportions of atoms that constitute a particular chemical compound, using the standard abbreviations for the chemical elements, and subscripts to indicate the number of atoms involved. For example, water is composed of two hydrogen atoms bonded to one oxygen atom: the chemical formula is H_2O.

A compound can be converted to a different chemical composition by interaction with a second chemical compound via a chemical reaction. In this process, bonds between atoms are broken in both of the interacting compounds, and then bonds are reformed so that new associations are made between atoms. Schematically, this reaction could be described as AB + CD --> AC + BD, where A, B, C, and D are each unique atoms; and AB, CD, AC, and BD are each unique compounds.

A chemical element bonded to an identical chemical element is not a chemical compound since only one element, not two different elements, is involved. Examples are the diatomic molecule hydrogen (H_2) and the polyatomic molecule sulfur (S_8).

Definitions

Any substance consisting of two or more different types of atoms (chemical elements) in a fixed proportion of its atoms (i.e., stoichiometry) can be termed a *chemical compound*; the concept is most readily understood when considering pure chemical substances. It follows from their being composed of fixed proportions of two or more types of atoms that chemical compounds can be converted, via chemical reaction, into compounds or substances each having fewer atoms. In the case of non-stoichiometric compounds, the proportions may be reproducible with regard to their preparation, and give fixed proportions of their component elements, but proportions that are not integral [e.g., for palladium hydride, PdH_x ($0.02 < x < 0.58$)]. Chemical compounds have a unique and defined chemical structure held together in a defined spatial arrangement by chemical bonds. Chemical compounds can be molecular compounds held together by covalent bonds, salts held together by ionic bonds, intermetallic compounds held together by metallic bonds, or the subset of chemical complexes that are held together by coordinate covalent bonds. Pure chemical elements are generally not considered chemical compounds, failing the two or more atom requirement, though they often consist of molecules composed of multiple atoms (such as in the diatomic molecule H_2, or the polyatomic molecule S_8, etc.).

There is varying and sometimes inconsistent nomenclature differentiating substances, which include truly non-stoichiometric examples, from chemical compounds, which require the fixed ratios. Many solid chemical substances—for example many silicate minerals—are chemical substances, but do not have simple formulae reflecting chemically bonding of elements to one another in fixed ratios; even so, these crystalline substances are often called "non-stoichiometric compounds". It may be argued that they are related to, rather than being chemical compounds, insofar as the variability in their compositions is often due to either the presence of foreign elements trapped within the crystal structure of an otherwise known true *chemical compound*, or due to perturbations in structure relative to the known compound that arise because of an excess of deficit of the constituent elements at places in its structure; such non-stoichiometric substances form most of the crust and mantle of the Earth. Other compounds regarded as chemically identical may have varying amounts of heavy or light isotopes of the constituent elements, which changes the ratio of elements by mass slightly.

Elementary Concepts

Characteristic properties of compounds include that *elements in a compound are present in a definite proportion.* For example, the molecule of the compound water is composed of hydrogen and oxygen in a ratio of 2:1. In addition, *compounds have a definite set of properties,* and the elements that comprise a compound do not retain their original properties. For example, hydrogen, which is combustible and non-supportive of combustion, combines with oxygen, which is non-combustible and supportive of combustion, to produce the compound water, which is non-combustible and non-supportive of combustion.

Comparison to Mixtures

The physical and chemical properties of compounds differ from those of their constituent elements. This is one of the main criteria that distinguish a compound from a mixture of elements or other substances—in general, a mixture's properties are closely related to, and depend on, the properties of its constituents. Another criterion that distinguishes a compound from a mixture is that constituents of a mixture can usually be separated by simple mechanical means, such as filtering, evaporation, or magnetic force, but components of a compound can be separated only by a chemical reaction. However, mixtures can be created by mechanical means alone, but a compound can be created (either from elements or from other compounds, or a combination of the two) only by a chemical reaction.

Some mixtures are so intimately combined that they have some properties similar to compounds and may easily be mistaken for compounds. One example is alloys. Alloys are made mechanically, most commonly by heating the constituent metals to a liquid state, mixing them thoroughly, and then cooling the mixture quickly so that the constituents are trapped in the base metal. Other examples of compound-like mixtures include intermetallic compounds and solutions of alkali metals in a liquid form of ammonia.

Formula

A chemical formula is a way of expressing information about the proportions of atoms that constitute a particular chemical compound, using a single line of chemical element symbols, numbers, and sometimes also other symbols, such as parentheses, dashes, brackets, commas and plus (+) and minus (−) signs.

Compounds may be described using formulas in various formats. For compounds that exist as molecules, the formula for the molecular unit is shown. For polymeric materials, such as minerals and many metal oxides, the empirical formula is normally given, e.g. NaCl for table salt.

The elements in a chemical formula are normally listed in a specific order, called the Hill system. In this system, the carbon atoms (if there are any) are usually listed first, any hydrogen atoms are listed next, and all other elements follow in alphabetical order. If the formula contains no carbon, then all of the elements, including hydrogen, are listed alphabetically. There are, however, several important exceptions to the normal rules. For ionic compounds, the positive ion is almost always listed first and the negative ion is listed second. For oxides, oxygen is usually listed last.

In general, organic acids follow the normal rules with C and H coming first in the formula. For example, the formula for trifluoroacetic acid is usually written as $C_2HF_3O_2$. More descriptive formulas can convey structural information, such as writing the formula for trifluoroacetic acid as CF_3CO_2H. On the other hand, the chemical formulas for most inorganic acids and bases are exceptions to the normal rules. They are written according to the rules for ionic compounds (positive first, negative second), but they also follow rules that emphasize their Arrhenius definitions. To be specific, the formula for most inorganic acids begins with hydrogen and the formula for most bases ends with the hydroxide ion (OH^-). Formulas for inorganic compounds do not often convey structural information, as illustrated by the common use of the formula H_2SO_4 for a molecule (sulfuric acid) that contains no H-S bonds. A more descriptive presentation would be $O_2S(OH)_2$, but it is almost never written this way.

Phases and Thermal Properties

Compounds may have several possible phases. All compounds can exist as solids, at least at low enough temperatures. Molecular compounds may also exist as liquids, gases, and, in some cases, even plasmas. All compounds decompose upon applying heat. The temperature at which such fragmentation occurs is often called the decomposition temperature. Decomposition temperatures are not sharp and depend on pressure, temperature, and the concentration of each species in the compound.

Intermolecular Force

Intermolecular forces (IMFs) are forces of attraction or repulsion which act between neighboring particles (atoms, molecules, or ions). They are weak compared to the intramolecular forces, the forces which keep a molecule together. For example the covalent bond, involving the sharing of electron pairs between atoms is much stronger than the forces present between the neighboring molecules. They are an essential part of force fields frequently used in molecular mechanics.

The investigation of intermolecular forces starts from macroscopic observations which point out the existence and action of forces at a molecular level. These observations include non-ideal-gas thermodynamic behavior reflected by virial coefficients, vapor pressure, viscosity, superficial tension and absorption data. ' The first reference to the nature of microscopic forces is found in Alexis Clairaut's work *Theorie de la Figure de la Terre*. Other scientists who have contributed to the investigation of microscopic forces include: Laplace, Gauss, Maxwell and Boltzmann.

Attractive intermolecular forces are considered by the following types:

- Ion-induced dipole forces
- Ion-dipole forces
- van der Waals forces (Keesom force, Debye force, and London dispersion force)

Information on intermolecular force is obtained by macroscopic measurements of properties like viscosity, PVT data. The link to microscopic aspects is given by virial coefficients and Lennard-Jones potentials.

Dipole-Dipole Interactions

Dipole-dipole interactions are electrostatic interactions between permanent dipoles in molecules. These interactions tend to align the molecules to increase attraction (reducing potential energy).

An example of a dipole-dipole interaction can be seen in hydrogen chloride (HCl): the positive end of a polar molecule will attract the negative end of the other molecule and influence its position. Polar molecules have a net attraction between them. Examples of polar molecules include hydrogen chloride (HCl) and chloroform ($CHCl_3$).

$$\overset{\delta+}{\ce{H}} - \overset{\delta-}{\ce{Cl}} \cdots \overset{\delta+}{\ce{H}} - \overset{\delta-}{\ce{Cl}}$$

Often molecules contain dipolar groups, but have no overall dipole moment. This occurs if there is symmetry within the molecule that causes the dipoles to cancel each other out. This occurs in molecules such as tetrachloromethane and carbon dioxide. Note that the dipole-dipole interaction between two individual atoms is usually zero, since atoms rarely carry a permanent dipole.

Ion-Dipole and Ion-Induced Dipole Forces

Ion-dipole and ion-induced dipole forces are similar to dipole-dipole and induced-dipole interactions but involve ions, instead of only polar and non-polar molecules. Ion-dipole and ion-induced dipole forces are stronger than dipole-dipole interactions because the charge of any ion is much greater than the charge of a dipole moment. Ion-dipole bonding is stronger than hydrogen bonding.

An ion-dipole force consists of an ion and a polar molecule interacting. They align so that the positive and negative groups are next to one another, allowing for maximum attraction.

An ion-induced dipole force consists of an ion and a non-polar molecule interacting. Like a dipole-induced dipole force, the charge of the ion causes distortion of the electron cloud on the non-polar molecule.

Hydrogen Bonding

A hydrogen bond is the attraction between the lone pair of an electronegative atom and a hydrogen atom that is bonded to either nitrogen, oxygen, or fluorine. The hydrogen bond is often described as a strong electrostatic dipole-dipole interaction. However, it also has some features of covalent bonding: it is directional, stronger than a van der Waals interaction, produces interatomic distances shorter than the sum of van der Waals radius, and usually involves a limited number of interaction partners, which can be interpreted as a kind of valence.

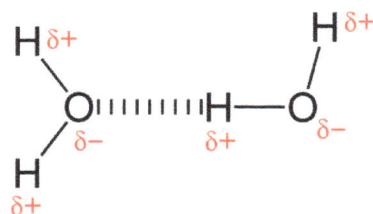

Intermolecular hydrogen bonding is responsible for the high boiling point of water (100 °C) compared to the other group 16 hydrides, which have no hydrogen bonds. Intramolecular hydrogen oxygen bonding is partly responsible for the secondary, tertiary, and quaternary structures of proteins and nucleic acids. It also plays an important role in the structure of polymers, both synthetic and natural.

Van Der Waals Forces

The van der Waals forces arise from interaction between uncharged atoms or molecules, leading not only to such phenomena as the cohesion of condensed phases and physical adsorption of gases, but also to a universal force of attraction between macroscopic bodies.

Keesom (Permanent-Permanent Dipoles) Interaction

The first contribution to van der Waals forces is due to electrostatic interactions between charges (in molecular ions), dipoles (for polar molecules), quadrupoles (all molecules with symmetry lower than cubic), and permanent multipoles. It is referred to as Keesom interactions(named after Willem Hendrik Keesom). These forces originate from the attraction between permanent dipoles (dipolar molecules) and are temperature dependent.

They consist of attractive interactions between dipoles that are ensemble averaged over different rotational orientations of the dipoles. It is assumed that the molecules are constantly rotating and never get locked into place. This is a good assumption, but at some point molecules do get locked into place. The energy of a Keesom interaction depends on the inverse sixth power of the distance, unlike the interaction energy of two spatially fixed dipoles, which depends on the inverse third power of the distance. The Keesom interaction can only occur among molecules that possess permanent dipole moments a.k.a. two polar molecules. Also Keesom interactions are very weak van der Waals interactions and do not occur in aqueous solutions that contain electrolytes. The angle averaged interaction is given by the following equation:

$$\frac{-2m_1^2 m_2^2}{48\pi^2 \varepsilon_o^2 \varepsilon_r^2 k_b T r^6} = V$$

Where m = charge per length, ε_o = permitivity of free space, ε_r = dielectric constant of surrounding material, T = temperature, k_b = Boltzmann constant, and r = distance between molecules.

Debye (Permanent-Induced Dipoles) Force

The second contribution is the induction (also known as polarization) or Debye force, arising from interactions between rotating permanent dipoles and from the polarizability of atoms and molecules (induced dipoles). These induced dipoles occur when one molecule with a permanent dipole repels another molecule's electrons. A molecule with permanent dipole can induce a dipole in a similar neighboring molecule and cause mutual attraction. Debye forces cannot occur between atoms. The forces between induced and permanent dipoles are not as temperature dependent as Keesom interactions because the induced dipole is free to shift and rotate around the non-polar molecule. The Debye induction effects and Keesom orientation effects are referred to as polar interactions.

The induced dipole forces appear from the induction (also known as polarization), which is the attractive interaction between a permanent multipole on one molecule with an induced (by the former di/multi-pole) multipole on another. This interaction is called the Debye force, named after Peter J.W. Debye.

One example of an induction-interaction between permanent dipole and induced dipole is the interaction between HCl and Ar. In this system, Ar experiences a dipole as its electrons are attracted (to the H side of HCl) or repelled (from the Cl side) by HCl. The angle averaged interaction is given by the following equation.

$$\frac{-m_1^2 \alpha_2}{16\pi^2 \varepsilon_o^2 \varepsilon_r^2 r^6} = V$$

Where   = polarizability

This kind of interaction can be expected between any polar molecule and non-polar/symmetrical molecule. The induction-interaction force is far weaker than dipole-dipole interaction, but stronger than the London dispersion force.

London Dispersion Force (Dipole-Induced Dipoles Interaction)

The third and dominant contribution is the dispersion or London force (fluctuating dipole-induced dipole), which arises due to the non-zero instantaneous dipole moments of all atoms and molecules. Such polarization can be induced either by a polar molecule or by the repulsion of negatively charged electron clouds in non-polar molecules. Thus, London interactions are caused by random fluctuations of electron density in an electron cloud. An atom with a large number of electrons will have a greater associated London force than an atom with fewer electrons. The dispersion (London) force is the most important component because all materials are polarizable, whereas Keesom and Debye forces require permanent dipoles. The London interaction is universal and is present in atom-atom interactions as well. For various reasons, London interactions (dispersion) have been considered relevant for interactions between macroscopic bodies in condensed systems. Hamaker developed the theory of van der Waals between macroscopic bodies in 1937 and showed that the additivity of these interactions renders them considerably more long-range.

Relative Strength of Forces

Bond type	Dissociation energy (kcal/mol)
Ionic Lattice Energy	250–4000
Covalent Bond Energy	30–260
Hydrogen Bonds	1–12 (about 5 in water)
Dipole–Dipole	0.5–2
London Dispersion Forces	<1 to 15 (estimated from the enthalpies of vaporization of hydrocarbons)

Note: this comparison is only approximate – the actual relative strengths will vary depending on the molecules involved. Ionic and covalent bonding will always be stronger than intermolecular forces in any given substance.

Effect on the Behavior of Gases

Intermolecular forces are repulsive at short distances and attractive at long distances. In a gas, the repulsive force chiefly has the effect of keeping two molecules from occupying the same volume. This gives a real gas a tendency to occupy a larger volume than an ideal gas at the same temperature and pressure. The attractive force draws molecules closer together and gives a real gas a tendency to occupy a smaller volume than an ideal gas. Which interaction is more important depends on temperature and pressure.

In a gas, the distances between molecules are generally large, so intermolecular forces have only a small effect. The attractive force is not overcome by the repulsive force, but by the thermal energy of the molecules. Temperature is the measure of thermal energy, so increasing temperature reduces the influence of the attractive force. In contrast, the influence of the repulsive force is essentially unaffected by temperature.

When a gas is compressed to increase its density, the influence of the attractive force increases. If the gas is made sufficiently dense, the attractions can become large enough to overcome the tendency of thermal motion to cause the molecules to spread out. Then the gas can condense to form a solid or liquid (i.e., a condensed phase). Lower temperature favors the formation of a condensed phase. In a condensed phase, there is very nearly a balance between the attractive and repulsive forces.

Quantum Mechanical Theories

Intermolecular forces observed between atoms and molecules can be described phenomenologically as occurring between permanent and instantaneous dipoles, as outlined above. Alternatively, one may seek a fundamental, unifying theory that is able to explain the various types of interactions such as hydrogen bonding, van der Waals forces and dipole-dipole interactions. Typically, this is done by applying the ideas of quantum mechanics to molecules, and Rayleigh–Schrödinger perturbation theory has been especially effective in this regard. When applied to existing quantum chemistry methods, such a quantum mechanical explanation of intermolecular interactions, this provides an array of approximate methods that can be used to analyze intermolecular interactions.

Chemical Reactions

A thermite reaction using iron(III) oxide. The sparks flying outwards are globules of molten iron trailing smoke in their wake.

A chemical reaction is a process that leads to the transformation of one set of chemical substances to another. Classically, chemical reactions encompass changes that only involve the positions of electrons in the forming and breaking of chemical bonds between atoms, with no change to the nuclei (no change to the elements present), and can often be described by a chemical equation. Nuclear chemistry is a sub-discipline of chemistry that involves the chemical reactions of unstable and radioactive elements where both electronic and nuclear changes may occur.

The substance (or substances) initially involved in a chemical reaction are called reactants or reagents. Chemical reactions are usually characterized by a chemical change, and they yield one or more products, which usually have properties different from the reactants. Reactions often consist of a sequence of individual sub-steps, the so-called elementary reactions, and the information on the precise course of action is part of the reaction mechanism. Chemical reactions are described with chemical equations, which symbolically present the starting materials, end products, and sometimes intermediate products and reaction conditions.

Chemical reactions happen at a characteristic reaction rate at a given temperature and chemical concentration. Typically, reaction rates increase with increasing temperature because there is more thermal energy available to reach the activation energy necessary for breaking bonds between atoms.

Reactions may proceed in the forward or reverse direction until they go to completion or reach equilibrium. Reactions that proceed in the forward direction to approach equilibrium are often described as spontaneous, requiring no input of free energy to go forward. Non-spontaneous reactions require input of free energy to go forward (examples include charging a battery by applying an external electrical power source, or photosynthesis driven by absorption of electromagnetic radiation in the form of sunlight).

Different chemical reactions are used in combinations during chemical synthesis in order to obtain a desired product. In biochemistry, a consecutive series of chemical reactions (where the product of one reaction is the reactant of the next reaction) form metabolic pathways. These reactions are often catalyzed by protein enzymes. Enzymes increase the rates of biochemical reactions, so that metabolic syntheses and decompositions impossible under ordinary conditions can occur at the temperatures and concentrations present within a cell.

The general concept of a chemical reaction has been extended to reactions between entities smaller than atoms, including nuclear reactions, radioactive decays, and reactions between elementary particles as described by quantum field theory.

History

Chemical reactions such as combustion in fire, fermentation and the reduction of ores to metals were known since antiquity. Initial theories of transformation of materials were developed by Greek philosophers, such as the Four-Element Theory of Empedocles stating that any substance is composed of the four basic elements – fire, water, air and earth. In the Middle Ages, chemical transformations were studied by Alchemists. They attempted, in particular, to convert lead into gold, for which purpose they used reactions of lead and lead-copper alloys with sulfur.

Antoine Lavoisier developed the theory of combustion as a chemical reaction with oxygen

The production of chemical substances that do not normally occur in nature has long been tried, such as the synthesis of sulfuric and nitric acids attributed to the controversial alchemist Jābir ibn Hayyān. The process involved heating of sulfate and nitrate minerals such as copper sulfate, alum and saltpeter. In the 17th century, Johann Rudolph Glauber produced hydrochloric acid and sodium sulfate by reacting sulfuric acid and sodium chloride. With the development of the lead chamber process in 1746 and the Leblanc process, allowing large-scale production of sulfuric acid and sodium carbonate, respectively, chemical reactions became implemented into the industry. Further optimization of sulfuric acid technology resulted in the contact process in the 1880s, and the Haber process was developed in 1909–1910 for ammonia synthesis.

From the 16th century, researchers including Jan Baptist van Helmont, Robert Boyle and Isaac Newton tried to establish theories of the experimentally observed chemical transformations. The phlogiston theory was proposed in 1667 by Johann Joachim Becher. It postulated the existence of a fire-like element called "phlogiston", which was contained within combustible bodies and released during combustion. This proved to be false in 1785 by Antoine Lavoisier who found the correct explanation of the combustion as reaction with oxygen from the air.

Joseph Louis Gay-Lussac recognized in 1808 that gases always react in a certain relationship with each other. Based on this idea and the atomic theory of John Dalton, Joseph Proust had developed the law of definite proportions, which later resulted in the concepts of stoichiometry and chemical equations.

Regarding the organic chemistry, it was long believed that compounds obtained from living organisms were too complex to be obtained synthetically. According to the concept of vitalism, organic matter was endowed with a "vital force" and distinguished from inorganic materials. This separation was ended however by the synthesis of urea from inorganic precursors by Friedrich Wöhler in 1828. Other chemists who brought major contributions to organic chemistry include Alexander William Williamson with his synthesis of ethers and Christopher Kelk Ingold, who, among many discoveries, established the mechanisms of substitution reactions.

Equations

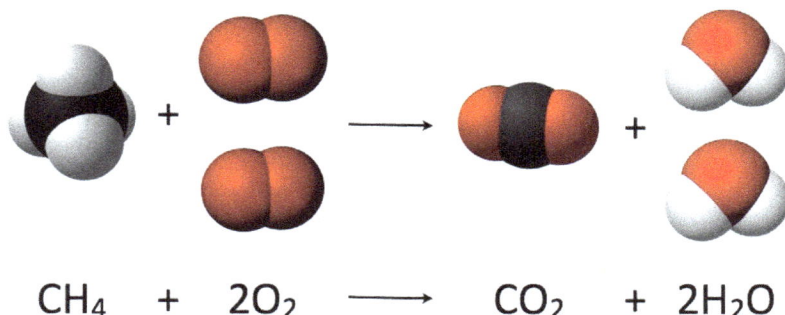

$$CH_4 \quad + \quad 2O_2 \quad \longrightarrow \quad CO_2 \quad + \quad 2H_2O$$

As seen from the equation $CH_4 + 2\,O_2 \rightarrow CO_2 + 2\,H_2O$, a coefficient of 2 must be placed before the oxygen gas on the reactants side and before the water on the products side in order for, as per the law of conservation of mass, the quantity of each element does not change during the reaction

Chemical equations are used to graphically illustrate chemical reactions. They consist of chemical or structural formulas of the reactants on the left and those of the products on the right. They are separated by an arrow ($\rightarrow$) which indicates the direction and type of the reaction; the arrow is read as the word "yields". The tip of the arrow points in the direction in which the reaction proceeds. A double arrow ($\rightleftharpoons$) pointing in opposite directions is used for equilibrium reactions. Equations should be balanced according to the stoichiometry, the number of atoms of each species should be the same on both sides of the equation. This is achieved by scaling the number of involved molecules (A, B, C and D in a schematic example below) by the appropriate integers a, b, c and d.

$$aA + bB -> cC + dD$$

More elaborate reactions are represented by reaction schemes, which in addition to starting materials and products show important intermediates or transition states. Also, some relatively minor additions to the reaction can be indicated above the reaction arrow; examples of such additions are water, heat, illumination, a catalyst, etc. Similarly, some minor products can be placed below the arrow, often with a minus sign.

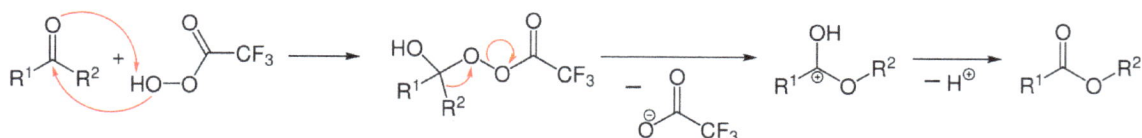

An example of organic reaction: oxidation of ketones to esters with a peroxycarboxylic acid

Retrosynthetic analysis can be applied to design a complex synthesis reaction. Here the analysis starts from the products, for example by splitting selected chemical bonds, to arrive at plausible initial reagents. A special arrow ($\Rightarrow$) is used in retro reactions.

Elementary Reactions

The elementary reaction is the smallest division into which a chemical reaction can be decomposed, it has no intermediate products. Most experimentally observed reactions are built up from many elementary reactions that occur in parallel or sequentially. The actual sequence of the individual elementary reactions is known as reaction mechanism. An elementary reaction involves a few molecules, usually one or two, because of the low probability for several molecules to meet at a certain time.

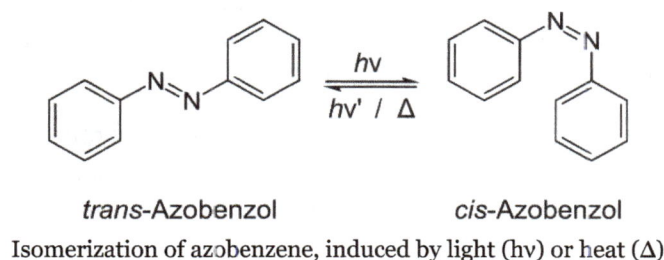

trans-Azobenzol   *cis*-Azobenzol

Isomerization of azobenzene, induced by light (hv) or heat (Δ)

The most important elementary reactions are unimolecular and bimolecular reactions. Only one molecule is involved in a unimolecular reaction; it is transformed by an isomerization or a dissociation into one or more other molecules. Such reactions require the addition of energy in the form of heat or light. A typical example of a unimolecular reaction is the cis–trans isomerization, in which the cis-form of a compound converts to the trans-form or vice versa.

In a typical dissociation reaction, a bond in a molecule splits (ruptures) resulting in two molecular fragments. The splitting can be homolytic or heterolytic. In the first case, the bond is divided so that each product retains an electron and becomes a neutral radical. In the second case, both electrons of the chemical bond remain with one of the products, resulting in charged ions. Dissociation plays an important role in triggering chain reactions, such as hydrogen–oxygen or polymerization reactions.

$AB -> A + B$

Dissociation of a molecule AB into fragments A and B

For bimolecular reactions, two molecules collide and react with each other. Their merger is called chemical synthesis or an addition reaction.

$A + B -> AB$

Another possibility is that only a portion of one molecule is transferred to the other molecule. This type of reaction occurs, for example, in redox and acid-base reactions. In redox reactions, the transferred particle is an electron, whereas in acid-base reactions it is a proton. This type of reaction is also called metathesis.

$HA + B -> A + HB$

for example

$NaCl + AgNO3 -> NaNO3 + AgCl(v)$

Chemical Equilibrium

Most chemical reactions are reversible, that is they can and do run in both directions. The forward and reverse reactions are competing with each other and differ in reaction rates. These rates depend on the concentration and therefore change with time of the reaction: the reverse rate gradually increases and becomes equal to the rate of the forward reaction, establishing the so-called chemical equilibrium. The time to reach equilibrium depends on such parameters as temperature, pressure and the materials involved, and is determined by the minimum free energy. In equilibrium, the Gibbs free energy must be zero. The pressure dependence can be explained with the Le Chatelier's principle. For example, an increase in pressure due to decreasing volume causes the reaction to shift to the side with the fewer moles of gas.

The reaction yield stabilizes at equilibrium, but can be increased by removing the product from the reaction mixture or changed by increasing the temperature or pressure. A change in the concentrations of the reactants does not affect the equilibrium constant, but does affect the equilibrium position.

Thermodynamics

Chemical reactions are determined by the laws of thermodynamics. Reactions can proceed by themselves if they are exergonic, that is if they release energy. The associated free energy of the reaction is composed of two different thermodynamic quantities, enthalpy and entropy:

$\Delta G = \Delta H - T \cdot \Delta S$.

G: free energy, H: enthalpy, T: temperature, S: entropy, Δ: difference(change between original and product)

Reactions can be exothermic, where ΔH is negative and energy is released. Typical examples of exothermic reactions are precipitation and crystallization, in which ordered solids are formed from disordered gaseous or liquid phases. In contrast, in endothermic reactions, heat is consumed from the environment. This can occur by increasing the entropy of the system, often through the formation of gaseous reaction products, which have high entropy. Since the entropy increases with temperature, many endothermic reactions preferably take place at high temperatures. On the contrary, many exothermic reactions such as crystallization occur at low temperatures. Changes in temperature can sometimes reverse the sign of the enthalpy of a reaction, as for the carbon monoxide reduction of molybdenum dioxide:

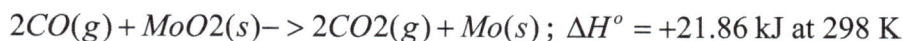

$2CO(g) + MoO2(s) -> 2CO2(g) + Mo(s)$; $\Delta H^{o} = +21.86$ kJ at 298 K

This reaction to form carbon dioxide and molybdenum is endothermic at low temperatures, becoming less so with increasing temperature. $\Delta H°$ is zero at 1855 K, and the reaction becomes exothermic above that temperature.

Changes in temperature can also reverse the direction tendency of a reaction. For example, the water gas shift reaction

$CO(g) + H2O(v) <=> CO2(g) + H2(g)$

is favored by low temperatures, but its reverse is favored by high temperature. The shift in reaction direction tendency occurs at 1100 K.

Reactions can also be characterized by the internal energy which takes into account changes in the entropy, volume and chemical potential. The latter depends, among other things, on the activities of the involved substances.

$$dU = T \cdot dS - p \cdot dV + \mu \cdot dn$$

U: internal energy, S: entropy, p: pressure, μ: chemical potential, n: number of molecules, d: small change sign

Kinetics

The speed at which reactions takes place is studied by reaction kinetics. The rate depends on various parameters, such as:

- Reactant concentrations, which usually make the reaction happen at a faster rate if raised through increased collisions per unit time. Some reactions, however, have rates that are *independent* of reactant concentrations. These are called zero order reactions.

- Surface area available for contact between the reactants, in particular solid ones in heterogeneous systems. Larger surface areas lead to higher reaction rates.

- Pressure – increasing the pressure decreases the volume between molecules and therefore increases the frequency of collisions between the molecules.

- Activation energy, which is defined as the amount of energy required to make the reaction start and carry on spontaneously. Higher activation energy implies that the reactants need more energy to start than a reaction with a lower activation energy.

- Temperature, which hastens reactions if raised, since higher temperature increases the energy of the molecules, creating more collisions per unit time,

- The presence or absence of a catalyst. Catalysts are substances which change the pathway (mechanism) of a reaction which in turn increases the speed of a reaction by lowering the activation energy needed for the reaction to take place. A catalyst is not destroyed or changed during a reaction, so it can be used again.

- For some reactions, the presence of electromagnetic radiation, most notably ultraviolet light, is needed to promote the breaking of bonds to start the reaction. This is particularly true for reactions involving radicals.

Several theories allow calculating the reaction rates at the molecular level. This field is referred to as reaction dynamics. The rate v of a first-order reaction, which could be disintegration of a substance A, is given by:

$$v = -\frac{d[A]}{dt} = k \cdot [A].$$

Its integration yields:

$$[A](t) = [A]_0 \cdot e^{-k \cdot t}.$$

Here k is first-order rate constant having dimension 1/time, [A](t) is concentration at a time t and $[A]_o$ is the initial concentration. The rate of a first-order reaction depends only on the concentration and the properties of the involved substance, and the reaction itself can be described with the characteristic half-life. More than one time constant is needed when describing reactions of higher order. The temperature dependence of the rate constant usually follows the Arrhenius equation:

$$k = k_0 e - E_a >$$

where E_a is the activation energy and k_B is the Boltzmann constant. One of the simplest models of reaction rate is the collision theory. More realistic models are tailored to a specific problem and include the transition state theory, the calculation of the potential energy surface, the Marcus theory and the Rice–Ramsperger–Kassel–Marcus (RRKM) theory.

Reaction Types

Four Basic Types

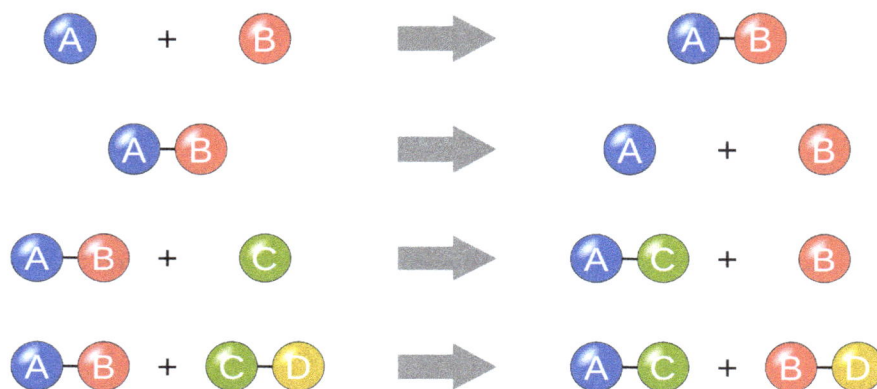

Representation of four basic chemical reactions types: synthesis, decomposition, single replacement and double replacement.

Synthesis

In a synthesis reaction, two or more simple substances combine to form a more complex substance. These reactions are in the general form:

$$A + B - > AB$$

Two or more reactants yielding one product is another way to identify a synthesis reaction. One example of a synthesis reaction is the combination of iron and sulfur to form iron(II) sulfide:

$$8Fe + S8 - > 8FeS$$

Another example is simple hydrogen gas combined with simple oxygen gas to produce a more complex substance, such as water.

Decomposition

A decomposition reaction is when a more complex substance breaks down into its more simple parts. It is thus the opposite of a synthesis reaction, and can be written as

$$AB -> A + B$$

One example of a decomposition reaction is the electrolysis of water to make oxygen and hydrogen gas:

$$2H2O -> 2H2 + O2$$

Single Replacement

In a single replacement reaction, a single uncombined element replaces another in a compound; in other words, one element trades places with another element in a compound These reactions come in the general form of:

$$A + BC -> AC + B$$

One example of a single displacement reaction is when magnesium replaces hydrogen in water to make magnesium hydroxide and hydrogen gas:

$$Mg + 2H2O -> Mg(OH)2 + H2 \uparrow$$

Double Replacement

In a double replacement reaction, the anions and cations of two compounds switch places and form two entirely different compounds. These reactions are in the general form:

$$AB + CD -> AD + CB$$

For example, when barium chloride ($BaCl_2$) and magnesium sulfate ($MgSO_4$) react, the SO_4^{2-} anion switches places with the $2Cl^-$ anion, giving the compounds $BaSO_4$ and $MgCl_2$.

Another example of a double displacement reaction is the reaction of lead(II) nitrate with potassium iodide to form lead(II) iodide and potassium nitrate:

$$Pb(NO3)2 + 2KI -> PbI2(v) + 2KNO3$$

Oxidation and Reduction

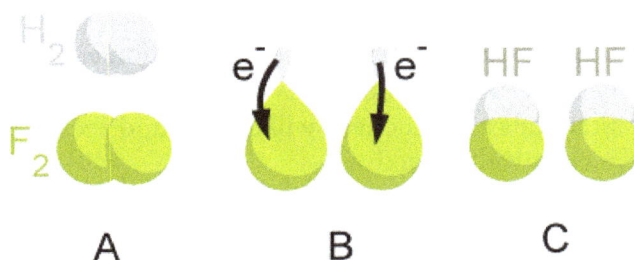

Illustration of a redox reaction

Sodium chloride is formed through the redox reaction of sodium metal and chlorine gas

Redox reactions can be understood in terms of transfer of electrons from one involved species (reducing agent) to another (oxidizing agent). In this process, the former species is *oxidized* and the latter is *reduced*. Though sufficient for many purposes, these descriptions are not precisely correct. Oxidation is better defined as an increase in oxidation state, and reduction as a decrease in oxidation state. In practice, the transfer of electrons will always change the oxidation state, but there are many reactions that are classed as "redox" even though no electron transfer occurs (such as those involving covalent bonds).

In the following redox reaction, hazardous sodium metal reacts with toxic chlorine gas to form the ionic compound sodium chloride, or common table salt:

$$2Na(s) + Cl2(g) -> 2NaCl(s)$$

In the reaction, sodium metal goes from an oxidation state of 0 (as it is a pure element) to +1: in other words, the sodium lost one electron and is said to have been oxidized. On the other hand, the chlorine gas goes from an oxidation of 0 (it is also a pure element) to −1: the chlorine gains one electron and is said to have been reduced. Because the chlorine is the one reduced, it is considered the electron acceptor, or in other words, induces oxidation in the sodium – thus the chlorine gas is considered the oxidizing agent. Conversely, the sodium is oxidized or is the electron donor, and thus induces reduction in the other species and is considered the *reducing agent*.

Which of the involved reactants would be reducing or oxidizing agent can be predicted from the electronegativity of their elements. Elements with low electronegativity, such as most metals, easily donate electrons and oxidize – they are reducing agents. On the contrary, many ions with high oxidation numbers, such as H2O2, MnO−4, CrO3, Cr2O2−7, OsO4 can gain one or two extra electrons and are strong oxidizing agents.

The number of electrons donated or accepted in a redox reaction can be predicted from the electron configuration of the reactant element. Elements try to reach the low-energy noble gas configuration, and therefore alkali metals and halogens will donate and accept one electron respectively. Noble gases themselves are chemically inactive.

An important class of redox reactions are the electrochemical reactions, where electrons from the power supply are used as the reducing agent. These reactions are particularly important for the production of chemical elements, such as chlorine or aluminium. The reverse process in which electrons are released in redox reactions and can be used as electrical energy is possible and used in batteries.

Complexation

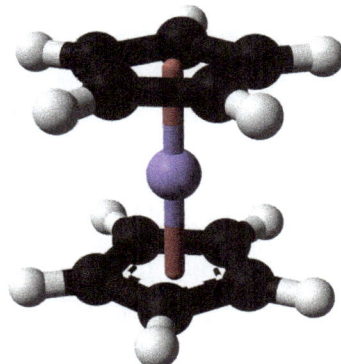

Ferrocene – an iron atom sandwiched between two C_5H_5 ligands

In complexation reactions, several ligands react with a metal atom to form a coordination complex. This is achieved by providing lone pairs of the ligand into empty orbitals of the metal atom and forming dipolar bonds. The ligands are Lewis bases, they can be both ions and neutral molecules, such as carbon monoxide, ammonia or water. The number of ligands that react with a central metal atom can be found using the 18-electron rule, saying that the valence shells of a transition metal will collectively accommodate 18 electrons, whereas the symmetry of the resulting complex can be predicted with the crystal field theory and ligand field theory. Complexation reactions also include ligand exchange, in which one or more ligands are replaced by another, and redox processes which change the oxidation state of the central metal atom.

Acid-Base Reactions

In the Brønsted–Lowry acid–base theory, an acid-base reaction involves a transfer of protons (H^+) from one species (the acid) to another (the base). When a proton is removed from an acid, the resulting species is termed that acid's conjugate base. When the proton is accepted by a base, the resulting species is termed that base's conjugate acid. In other words, acids act as proton donors and bases act as proton acceptors according to the following equation:

$$\underset{acid}{HA} + \underset{base}{B} \iff \underset{conjugated\ base}{A^-} + \underset{conjugated\ acid}{HB+}$$

The reverse reaction is possible, and thus the acid/base and conjugated base/acid are always in equilibrium. The equilibrium is determined by the acid and base dissociation constants (K_a and K_b) of the involved substances. A special case of the acid-base reaction is the neutralization where an acid and a base, taken at exactly same amounts, form a neutral salt.

Acid-base reactions can have different definitions depending on the acid-base concept employed. Some of the most common are:

- Arrhenius definition: Acids dissociate in water releasing H_3O^+ ions; bases dissociate in water releasing OH^- ions.

- Brønsted-Lowry definition: Acids are proton (H⁺) donors, bases are proton acceptors; this includes the Arrhenius definition.

- Lewis definition: Acids are electron-pair acceptors, bases are electron-pair donors; this includes the Brønsted-Lowry definition.

Precipitation

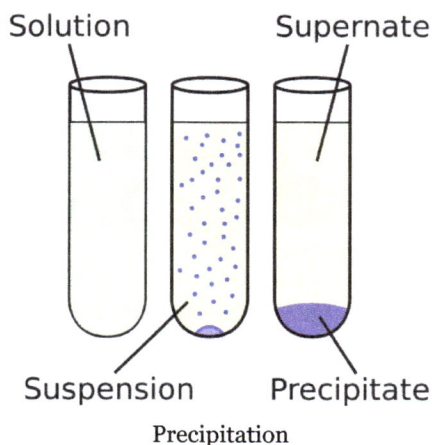

Precipitation

Precipitation is the formation of a solid in a solution or inside another solid during a chemical reaction. It usually takes place when the concentration of dissolved ions exceeds the solubility limit and forms an insoluble salt. This process can be assisted by adding a precipitating agent or by removal of the solvent. Rapid precipitation results in an amorphous or microcrystalline residue and slow process can yield single crystals. The latter can also be obtained by recrystallization from microcrystalline salts.

Solid-State Reactions

Reactions can take place between two solids. However, because of the relatively small diffusion rates in solids, the corresponding chemical reactions are very slow in comparison to liquid and gas phase reactions. They are accelerated by increasing the reaction temperature and finely dividing the reactant to increase the contacting surface area.

Reactions at the solid|Gas Interface

Reaction can take place at the solid|gas interface, surfaces at very low pressure such as ultra-high vacuum. Via scanning tunneling microscopy, it is possible to observe reactions at the solid|gas interface in real space, if the time scale of the reaction is in the correct range. Reactions at the solid|gas interface are in some cases related to catalysis.

Photochemical Reactions

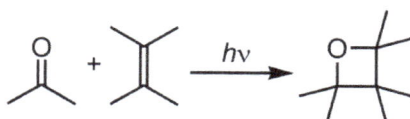

In this Paterno–Büchi reaction, a photoexcited carbonyl group is added to an unexcited olefin, yielding an oxetane.

In photochemical reactions, atoms and molecules absorb energy (photons) of the illumination light and convert into an excited state. They can then release this energy by breaking chemical bonds, thereby producing radicals. Photochemical reactions include hydrogen–oxygen reactions, radical polymerization, chain reactions and rearrangement reactions.

Many important processes involve photochemistry. The premier example is photosynthesis, in which most plants use solar energy to convert carbon dioxide and water into glucose, disposing of oxygen as a side-product. Humans rely on photochemistry for the formation of vitamin D, and vision is initiated by a photochemical reaction of rhodopsin. In fireflies, an enzyme in the abdomen catalyzes a reaction that results in bioluminescence. Many significant photochemical reactions, such as ozone formation, occur in the Earth atmosphere and constitute atmospheric chemistry.

Catalysis

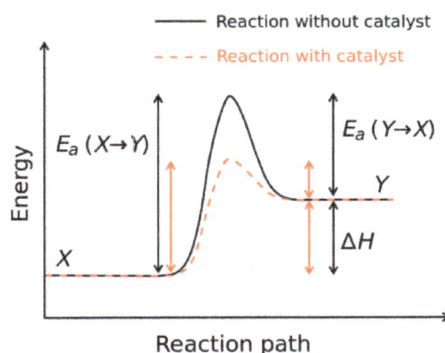

Schematic potential energy diagram showing the effect of a catalyst in an endothermic chemical reaction. The presence of a catalyst opens a different reaction pathway (in red) with a lower activation energy. The final result and the overall thermodynamics are the same.

Solid heterogeneous catalysts are plated on meshes in ceramic catalytic converters in order to maximize their surface area. This exhaust converter is from a Peugeot 106 S2 1100

In catalysis, the reaction does not proceed directly, but through reaction with a third substance known as catalyst. Although the catalyst takes part in the reaction, it is returned to its original state by the end of the reaction and so is not consumed. However, it can be inhibited, deactivated or destroyed by secondary processes. Catalysts can be used in a different phase (heterogeneous) or in the same phase (homogeneous) as the reactants. In heterogeneous catalysis, typical secondary processes include coking where the catalyst becomes covered by polymeric side products. Additionally, heterogeneous catalysts can dissolve into the solution in a solid–liquid system or evaporate in a solid–gas system. Catalysts can only speed up the reaction – chemicals that slow down the reaction are called inhibitors. Substances that increase the activity of catalysts are called promoters, and substances that deactivate catalysts are called catalytic poisons. With a catalyst, a reaction which is kinetically inhibited by a high activation energy can take place in circumvention of this activation energy.

Heterogeneous catalysts are usually solids, powdered in order to maximize their surface area. Of particular importance in heterogeneous catalysis are the platinum group metals and other transition metals, which are used in hydrogenations, catalytic reforming and in the synthesis of commodity chemicals such as nitric acid and ammonia. Acids are an example of a homogeneous catalyst, they increase the nucleophilicity of carbonyls, allowing a reaction that would not otherwise proceed with electrophiles. The advantage of homogeneous catalysts is the ease of mixing them with the reactants, but they may also be difficult to separate from the products. Therefore, heterogeneous catalysts are preferred in many industrial processes.

Reactions in Organic Chemistry

In organic chemistry, in addition to oxidation, reduction or acid-base reactions, a number of other reactions can take place which involve covalent bonds between carbon atoms or carbon and heteroatoms (such as oxygen, nitrogen, halogens, etc.). Many specific reactions in organic chemistry are name reactions designated after their discoverers.

Substitution

In a substitution reaction, a functional group in a particular chemical compound is replaced by another group. These reactions can be distinguished by the type of substituting species into a nucleophilic, electrophilic or radical substitution.

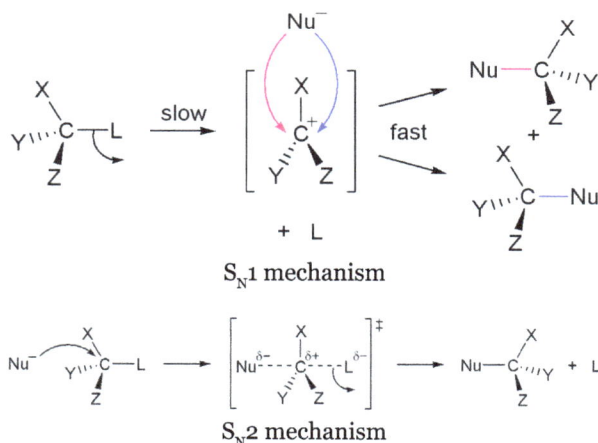

S_N1 mechanism

S_N2 mechanism

In the first type, a nucleophile, an atom or molecule with an excess of electrons and thus a negative charge or partial charge, replaces another atom or part of the "substrate" molecule. The electron pair from the nucleophile attacks the substrate forming a new bond, while the leaving group departs with an electron pair. The nucleophile may be electrically neutral or negatively charged, whereas the substrate is typically neutral or positively charged. Examples of nucleophiles are hydroxide ion, alkoxides, amines and halides. This type of reaction is found mainly in aliphatic hydrocarbons, and rarely in aromatic hydrocarbon. The latter have high electron density and enter nucleophilic aromatic substitution only with very strong electron withdrawing groups. Nucleophilic substitution can take place by two different mechanisms, S_N1 and S_N2. In their names, S stands for substitution, N for nucleophilic, and the number represents the kinetic order of the reaction, unimolecular or bimolecular.

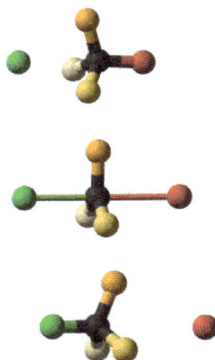

The three steps of an S_N2 reaction. The nucleophile is green and the leaving group is red

S_N2 reaction causes stereo inversion (Walden inversion)

The S_N1 reaction proceeds in two steps. First, the leaving group is eliminated creating a carbocation. This is followed by a rapid reaction with the nucleophile.

In the S_N2 mechanism, the nucleophile forms a transition state with the attacked molecule, and only then the leaving group is cleaved. These two mechanisms differ in the stereochemistry of the products. S_N1 leads to the non-stereospecific addition and does not result in a chiral center, but rather in a set of geometric isomers (*cis/trans*). In contrast, a reversal (Walden inversion) of the previously existing stereochemistry is observed in the S_N2 mechanism.

Electrophilic substitution is the counterpart of the nucleophilic substitution in that the attacking atom or molecule, an electrophile, has low electron density and thus a positive charge. Typical electrophiles are the carbon atom of carbonyl groups, carbocations or sulfur or nitronium cations. This reaction takes place almost exclusively in aromatic hydrocarbons, where it is called electrophilic aromatic substitution. The electrophile attack results in the so-called σ-complex, a transition state in which the aromatic system is abolished. Then, the leaving group, usually a proton, is split off and the aromaticity is restored. An alternative to aromatic substitution is electrophilic aliphatic substitution. It is similar to the nucleophilic aliphatic substitution and also has two major types, S_E1 and S_E2

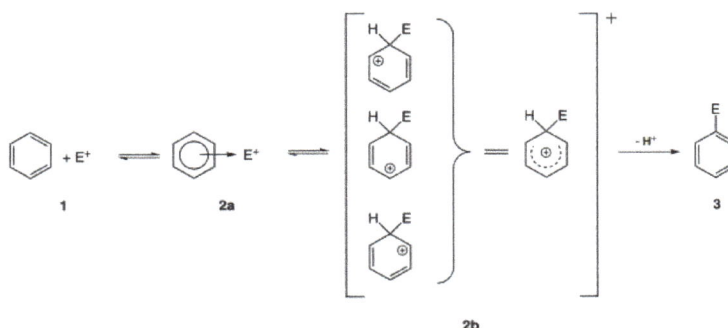

Mechanism of electrophilic aromatic substitution

In the third type of substitution reaction, radical substitution, the attacking particle is a radical. This process usually takes the form of a chain reaction, for example in the reaction of alkanes with halogens. In the first step, light or heat disintegrates the halogen-containing molecules producing the radicals. Then the reaction proceeds as an avalanche until two radicals meet and recombine.

$$X. + R - H -> X - H + R.$$

$$R. + X2 -> R - X + X.$$

Reactions during the chain reaction of radical substitution

Addition and Elimination

The addition and its counterpart, the elimination, are reactions which change the number of substitutents on the carbon atom, and form or cleave multiple bonds. Double and triple bonds can be produced by eliminating a suitable leaving group. Similar to the nucleophilic substitution, there are several possible reaction mechanisms which are named after the respective reaction order. In the E1 mechanism, the leaving group is ejected first, forming a carbocation. The next step, formation of the double bond, takes place with elimination of a proton (deprotonation). The leaving order is reversed in the E1cb mechanism, that is the proton is split off first. This mechanism requires participation of a base. Because of the similar conditions, both reactions in the E1 or E1cb elimination always compete with the S_N1 substitution.

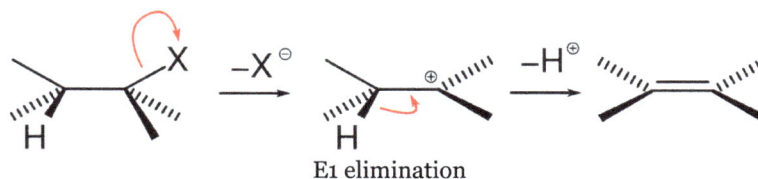

E1 elimination

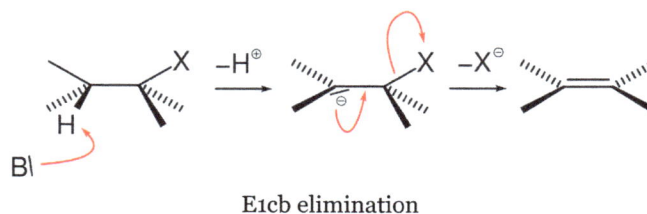

E1cb elimination

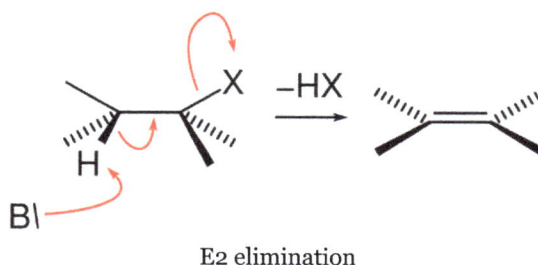

E2 elimination

The E2 mechanism also requires a base, but there the attack of the base and the elimination of the leaving group proceed simultaneously and produce no ionic intermediate. In contrast to the E1 eliminations, different stereochemical configurations are possible for the reaction product in the E2 mechanism, because the attack of the base preferentially occurs in the anti-position with respect to the leaving group. Because of the similar conditions and reagents, the E2 elimination is always in competition with the S_N2-substitution.

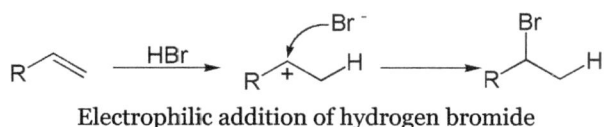

Electrophilic addition of hydrogen bromide

The counterpart of elimination is the addition where double or triple bonds are converted into single bonds. Similar to the substitution reactions, there are several types of additions distinguished by the type of the attacking particle. For example, in the electrophilic addition of hydrogen bromide, an electrophile (proton) attacks the double bond forming a carbocation, which then reacts with the nucleophile (bromine). The carbocation can be formed on either side of the double bond depending on the groups attached to its ends, and the preferred configuration can be predicted with the Markovnikov's rule. This rule states that "In the heterolytic addition of a polar molecule to an alkene or alkyne, the more electronegative (nucleophilic) atom (or part) of the polar molecule becomes attached to the carbon atom bearing the smaller number of hydrogen atoms."

If the addition of a functional group takes place at the less substituted carbon atom of the double bond, then the electrophilic substitution with acids is not possible. In this case, one has to use the hydroboration–oxidation reaction, where in the first step, the boron atom acts as electrophile and adds to the less substituted carbon atom. At the second step, the nucleophilic hydroperoxide or halogen anion attacks the boron atom.

While the addition to the electron-rich alkenes and alkynes is mainly electrophilic, the nucleophilic addition plays an important role for the carbon-heteroatom multiple bonds, and especially its most important representative, the carbonyl group. This process is often associated with an elimination, so that after the reaction the carbonyl group is present again. It is therefore called addition-elimination reaction and may occur in carboxylic acid derivatives such as chlorides, esters or anhydrides. This reaction is often catalyzed by acids or bases, where the acids increase by the electrophilicity of the carbonyl group by binding to the oxygen atom, whereas the bases enhance the nucleophilicity of the attacking nucleophile.

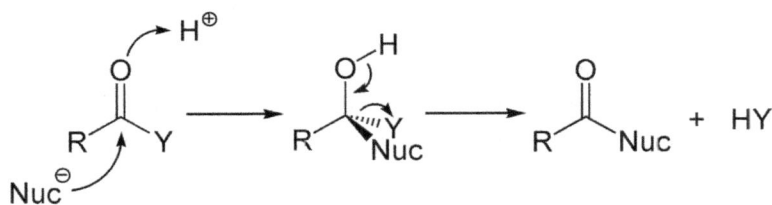

Acid-catalyzed addition-elimination mechanism

Nucleophilic addition of a carbanion or another nucleophile to the double bond of an alpha, beta unsaturated carbonyl compound can proceed via the Michael reaction, which belongs to the larger class of conjugate additions. This is one of the most useful methods for the mild formation of C–C bonds.

Some additions which can not be executed with nucleophiles and electrophiles, can be succeeded with free radicals. As with the free-radical substitution, the radical addition proceeds as a chain reaction, and such reactions are the basis of the free-radical polymerization.

Other Organic Reaction Mechanisms

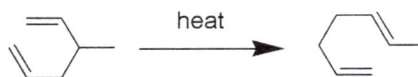

The Cope rearrangement of 3-methyl-1,5-hexadiene

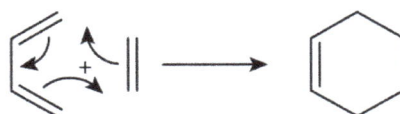

Mechanism of a Diels-Alder reaction

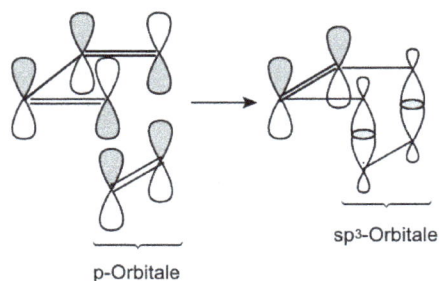

Orbital overlap in a Diels-Alder reaction

In a rearrangement reaction, the carbon skeleton of a molecule is rearranged to give a structural isomer of the original molecule. These include hydride shift reactions such as the Wagner-Meerwein rearrangement, where a hydrogen, alkyl or aryl group migrates from one carbon to a neighboring carbon. Most rearrangements are associated with the breaking and formation of new carbon-carbon bonds. Other examples are sigmatropic reaction such as the Cope rearrangement.

Cyclic rearrangements include cycloadditions and, more generally, pericyclic reactions, wherein two or more double bond-containing molecules form a cyclic molecule. An important example of cycloaddition reaction is the Diels–Alder reaction (the so-called [4+2] cycloaddition) between a conjugated diene and a substituted alkene to form a substituted cyclohexene system.

Whether a certain cycloaddition would proceed depends on the electronic orbitals of the participating species, as only orbitals with the same sign of wave function will overlap and interact constructively to form new bonds. Cycloaddition is usually assisted by light or heat. These perturbations result in different arrangement of electrons in the excited state of the involved molecules and therefore in different effects. For example, the [4+2] Diels-Alder reactions can be assisted by heat whereas the [2+2] cycloaddition is selectively induced by light. Because of the orbital character, the potential for developing stereoisomeric products upon cycloaddition is limited, as described by the Woodward–Hoffmann rules.

Biochemical Reactions

Biochemical reactions are mainly controlled by enzymes. These proteins can specifically catalyze a single reaction, so that reactions can be controlled very precisely. The reaction takes place in the active site, a small part of the enzyme which is usually found in a cleft or pocket lined by amino acid residues, and the rest of the enzyme is used mainly for stabilization. The catalytic action of enzymes relies on several mechanisms including the molecular shape ("induced fit"), bond strain, proximity and orientation of molecules relative to the enzyme, proton donation or withdrawal (acid/base catalysis), electrostatic interactions and many others.

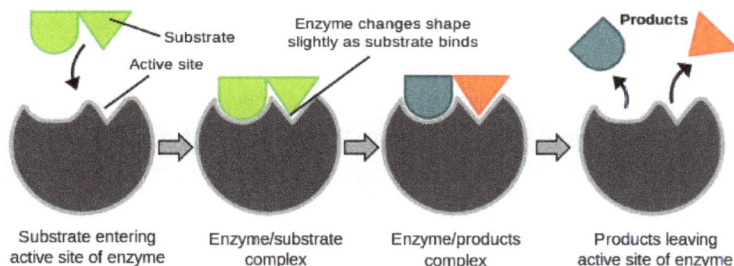

Illustration of the induced fit model of enzyme activity

The biochemical reactions that occur in living organisms are collectively known as metabolism. Among the most important of its mechanisms is the anabolism, in which different DNA and enzyme-controlled processes result in the production of large molecules such as proteins and carbohydrates from smaller units. Bioenergetics studies the sources of energy for such reactions. An important energy source is glucose, which can be produced by plants via photosynthesis or assimilated from food. All organisms use this energy to produce adenosine triphosphate (ATP), which can then be used to energize other reactions.

Applications

Thermite reaction proceeding in railway welding. Shortly after this, the liquid iron flows into the mould around the rail gap

Chemical reactions are central to chemical engineering where they are used for the synthesis of new compounds from natural raw materials such as petroleum and mineral ores. It is essential to make the reaction as efficient as possible, maximizing the yield and minimizing the amount of reagents, energy inputs and waste. Catalysts are especially helpful for reducing the energy required for the reaction and increasing its reaction rate.

Some specific reactions have their niche applications. For example, the thermite reaction is used to generate light and heat in pyrotechnics and welding. Although it is less controllable than the more conventional oxy-fuel welding, arc welding and flash welding, it requires much less equipment and is still used to mend rails, especially in remote areas.

Monitoring

Mechanisms of monitoring chemical reactions depend strongly on the reaction rate. Relatively slow processes can be analyzed in situ for the concentrations and identities of the individual ingredients. Important tools of real time analysis are the measurement of pH and analysis of optical absorption (color) and emission spectra. A less accessible but rather efficient method is introduction of a radioactive isotope into the reaction and monitoring how it changes over time and where it moves to; this method is often used to analyze redistribution of substances in the human body. Faster reactions are usually studied with ultrafast laser spectroscopy where utilization of femtosecond lasers allows short-lived transition states to be monitored at time scaled down to a few femtoseconds.

References

- *Atkins, Peter; Loretta Jones (1997). Chemistry: Molecules, Matter and Change. New York: W. H. Freeman & Co. pp. 294–295. ISBN 0-7167-3107-X.*

- *Whitten, Kenneth W.; Davis, Raymond E.; Peck, M. Larry (2000), General Chemistry (6th ed.), Fort Worth, TX: Saunders College Publishing/Harcourt College Publishers, ISBN 978-0-03-072373-5*

- *Brown, Theodore L.; LeMay, H. Eugene; Bursten, Bruce E.; Murphy, Catherine J.; Woodward, Patrick (2009), Chemistry: The Central Science, AP Edition (11th ed.), Upper Saddle River, NJ: Pearson/Prentice Hall, pp. 5–6, ISBN 0-13-236489-1*

- *Hill, John W.; Petrucci, Ralph H.; McCreary, Terry W.; Perry, Scott S. (2005), General Chemistry (4th ed.), Upper Saddle River, NJ: Pearson/Prentice Hall, p. 6, ISBN 978-0-13-140283-6*

- *Wilbraham, Antony; Matta, Michael; Staley, Dennis; Waterman, Edward (2002), Chemistry (1st ed.), Upper Saddle River, NJ: Pearson/Prentice Hall, p. 36, ISBN 0-13-251210-6*

- Margenau, H. and Kestner, N. (1969) *Theory of intermolecular forces*, International Series of Monographs in Natural Philosophy, Pergamon Press, ISBN 1483119289

- Majer, V. and Svoboda, V. (1985) *Enthalpies of Vaporization of Organic Compounds*, Blackwell Scientific Publications, Oxford. ISBN 0632015292

- *Kandori, Hideki (2006). "Retinal Binding Proteins". In Dugave, Christophe. Cis-trans Isomerization in Biochemistry. Wiley-VCH. p. 56. ISBN 3-527-31304-4.*

- *Guo, Liang-Hong; Allen, H.; Hill, O. (1991). "Direct Electrochemistry of Proteins and Enzymes". In A. G. Sykes. Advances in Inorganic Chemistry 36. San Diego: Academic Press. p. 359. ISBN 0-12-023636-2.*

- *Meyer, H. Jürgen (2007). "Festkörperchemie". In Erwin Riedel. Modern Inorganic Chemistry (in German) (3rd ed.). de Gruyter. p. 171. ISBN 978-3-11-019060-1.*

- *Elschenbroich, Christoph (2008). Organometallchemie (6th ed.). Wiesbaden: Vieweg+Teubner Verlag. p. 263. ISBN 978-3-8351-0167-8.*

- *March, Jerry (1985), Advanced Organic Chemistry: Reactions, Mechanisms, and Structure (3rd ed.), New York: Wiley, ISBN 0-471-85472-7*

- *Latscha, Hans Peter; Kazmaier, Uli; Klein, Helmut Alfons (2008). Organische Chemie: Chemie-basiswissen II (in German) 2 (6th ed.). Springer. p. 273. ISBN 978-3-540-77106-7.*

Permissions

Index

www.ingramcontent.com/pod-product-compliance
Lightning Source LLC
Chambersburg PA
CBHW061256190326
41458CB00011B/3688